D1566176

Science Policy Up Close

Science Policy Up Close

John H. Marburger III

Edited by
Robert P. Crease

Harvard University Press

Cambridge, Massachusetts
London, England
2015

Copyright © 2015 by the President and Fellows of Harvard College
All rights reserved
Printed in the United States of America

First printing

Library of Congress Cataloging-in-Publication Data

Marburger, John H. (John Harmen), III, 1941–2011, author.
Science policy up close / John H. Marburger III ; edited by Robert P. Crease.
pages   cm
Includes index.
ISBN 978-0-674-41709-0
1. Science and state—United States.   2. Science—Political aspects—
United States.   3. Marburger, John H. (John Harmen), III, 1941–2011.
4. Science consultants—United States—Biography.   5. Physicists—
United States—Biography.   I. Crease, Robert P., editor.   II. Title.
Q127.U6M277 2015
338.973'06—dc23   2014015891

Contents

Preface	*vii*
Introduction *by Robert P. Crease*	*1*
1. The Shoreham Commission	*12*
2. The Superconducting Super Collider and the Collider Decade	*30*
3. Managing a National Laboratory	*73*
4. Presidential Science Advisor I: Advice and Advocacy in Washington	*119*
5. Presidential Science Advisor II: Measuring and Prioritizing	*156*
6. The Science and Art of Science Policy	*207*
Acknowledgments	*239*
Index	*241*

Preface

Science is for me a way of life. More than a profession or a career, deeper than any other driving factor in my thoughts and choices, science has been my constant companion since childhood, expanding in my comprehension eventually to include my entire sphere of action. Discovery was never my goal; rather, I had an irresistible compulsion to get to the bottom of things—all things. I was stunned as a child by the actuality of my surroundings, and by the age of five I knew the word *science* and somehow felt how important it would be to me. Psychology must have a name for this. A touch of autism, or something like it, said a friend much later. My mother was a charismatic listener, nonjudgmental and sympathetic; my father, a self-taught pragmatic engineer. Both were fair-minded and objective in their regard for others, and I presume my own personality emerged from theirs. This combination of personal and intellectual qualities has served me well in a long and unusually coherent career in science and in academic and public service. In this book I trace one prominent theme in that career that may have more general interest than the other themes of research and academic administration that ordered my life.

The theme is policy in action. From 1980, when I became president of Stony Brook University, I was drawn into a sequence of major science policy issues, not as much as a policy maker as a policy actor. My ability to deal with people in an objective and productive way made it difficult to avoid early executive responsibilities. At the University of Southern

California, where I joined the departments of physics and electrical engineering in 1966, I was chairman of physics at the age of thirty-two and dean at thirty-five, and I was president of Stony Brook at age thirty-nine. These visible positions attracted requests for public service, and I found it difficult to say no. The closing of the Shoreham Nuclear Power Plant, the rise and fall of the Superconducting Super Collider, the trauma of Brookhaven National Laboratory, and nearly eight years as science advisor to President George W. Bush brought me up close to how science policy actually plays out in the real world.

Despite the huge appeal of "doing physics" and my reluctance to dilute my concentration, I was fascinated by the deep puzzles of human behavior. I saw them as part of the Nature to whose explication I was addicted, and their very existence stirred a kind of internal competitiveness: Could I understand what is going on here? Can I use this knowledge to make the human machinery work better? It was difficult to distinguish the attraction of these human challenges from those of the physics work I was doing, and I did both as long as I could. Policy work for me remained on the map of science, where I felt at home. My goal was never power or public recognition but a kind of inner pride of craftsmanship and accomplishment.

Science policy making and implementation are messy, ill-defined processes that are dominated by their specific contexts. The structure of U.S. governments (state and local, as well as federal), and especially their open and democratic nature; the uncertainties and incompleteness of science relevant to societal issues; and the conflicting aims of the major policy players conspire to undermine the coherence of policy in action. And yet in each situation, specific strategies and tactics exist that can foster a positive outcome. In the events described in this book, I attempted to understand the complexities of each situation and not simply reject them as chaos or escape them with rhetoric. In each case, I tried to match my behavior to the needs of the diverse situations in which I found myself. These efforts and their results may contain some ideas useful to others who care as much as I do about the value of rationality in public affairs.

—*John H. Marburger III (2011)*

Introduction

by Robert P. Crease

John H. Marburger III, the longest-serving U.S. presidential science advisor, was a skilled science administrator who had a fresh and unique approach to science policy and science leadership. He understood the new challenges faced by science administrators and policy makers in the changing social and political climate of the late twentieth and early twenty-first centuries. This book contains his reflections on how to cope with those challenges, gleaned from his experiences throughout what he liked to call a "long and unusually coherent career." Its landmarks include serving as president of Stony Brook University (1980–1994), chairman of the Universities Research Association overseeing the Superconducting Super Collider (1988–1994), director of Brookhaven National Laboratory (1998–2001), science advisor to President George W. Bush and director of the White House Office of Science and Technology Policy (2001–2009), and vice president for research at Stony Brook (2010–2011).

Born in 1941, Marburger was first captivated by science and engineering at the age of five, when his mother read him the Childcraft volume on *Science and Industry,* whose pictures illustrated the manufacturing of such things as steel and Portland cement. Obtaining his

BA from Princeton University in 1962 and PhD from Stanford University in 1967, he joined the physics department at the University of Southern California in 1966. After becoming chair of the physics department in 1972, he hosted an early morning TV show, *Frontiers of Electronics*. That experience gave him his first lesson in how to present science, and himself, to large audiences. Its high point came on February 21, 1973, when a magnitude 5.3 earthquake struck shortly after 6 a.m., awakening people throughout the San Fernando Valley. Their first instinct was to turn on the TV, where many encountered the prerecorded Marburger talking captivatingly about information theory. Later in life he would proudly say, "That episode had a *huge* audience!" Disasters, it seems, gave Marburger his biggest audiences and brought out the best in him.

Marburger's experience hosting a television show provided him with important lessons about how to present himself in an engaging and authoritative way. This extended down to his choice of clothes. "Style is an important form of communication," a newspaper article once quoted him as saying. "I'm conscious in the morning when I get dressed of what my clothing will say to people I meet. I always look at the calendar to see who I'll meet. I don't want my clothes to intrude. I try to make my clothes say I'm serious about what we're going to talk about."[1] Later, as director of Brookhaven National Laboratory, Marburger would startle the lab's photographers and video makers by making knowledgeable suggestions on how best to film him.

In 1976, Marburger became dean of USC's College of Letters, Arts, and Sciences. Articulate and responsible, he was made the administration's spokesman in crises, starting with a scandal over preferential treatment for athletes. These tasks forced him to address wider audiences, including nonscientists and antagonists. The role also taught him how to become the public face of an institution under fire without compromising his own integrity, and how to bear the burden of that representation when he as an individual was inescapably perceived as standing in for the institution.

---

[1] "A Matter of Style, Who's Hot," *Three Village Herald (Stony Brook)*, February 22, 1984.

INTRODUCTION   3

In 1980, age thirty-nine, Marburger became president of Stony Brook University on Long Island, not far from New York City. This was a new challenge: how to handle an institution full of advocates for departments, schools, and offices. He did this by trying to cultivate a broader climate in which the whole would flourish, and although each advocate might not be fully satisfied they would at least be familiar with the need of the whole. In this capacity he came to appreciate the difference between an *advocate*, someone who promotes a cause, and an *advisor*, someone aware of the need for the whole to flourish and therefore the need to balance attention and resources given to causes, however worthy in themselves. Marburger remained impressed by the significance of the distinction between advocacy and advice for the rest of his career, and he often invokes it in these writings. In his eyes, one defect of science policy, which he tried to rectify, is that it is dominated by advocates and lacks not only advisors but also suitable input to guide them.

Shortly after arriving at Stony Brook University, Marburger wrote a series of memos to himself titled "On Management," capturing his early reflections on the subject. In one, he wrote about the different resources available to a complex organization such as a university or laboratory, and what happens when these become in short supply:

> When there is not enough money to go around, we decide which are the most important needs and then spend to satisfy them, deferring the lesser needs until they grow to achieve higher priority. The same might be said of our use of any scarce resource, including space or energy. If the scarcity is well documented, and the prioritization conducted openly and objectively, then an enlightened community usually accepts the cycle of deprivation and fulfillment as natural.

Marburger then points to another kind of important resource that works in a different way: what he calls "executive time," or a manager's attention to a problem area. When executive time is withheld from a problem area, he writes, even enlightened communities find it hard to accept the deprivation as natural, and organizations often react in a way that worsens the problem and even damages its ability to function.

Executive time is a tangible resource that, at least theoretically, can be allocated as deliberately as other resources. Its expenditure is not easily documented, however, and it is much more difficult to convince the community, and indeed one's own executive colleagues, that the allocation is rational or complete. Furthermore, the criteria according to which one judges a problem to warrant more attention are not easily identified. Everyone is familiar with methods to estimate needs for money, space or energy, but the need for executive time (and its associated indirect cost) is difficult to assess a priori.

Marburger's memo continues by discussing the inevitability of problems caused by lack of executive time, the different ways these problems can evolve, the obstacles to solving them, and how they might best be addressed. He concludes, "I believe that the manner in which they [problems caused by inadequate executive time] are addressed gauges the management maturity of an organization."[2]

In the chapters that follow, the elements of Marburger's perspective on science policy appear not only in his explicit statements but also in how he chose to devote his executive time. For this reason, it is often difficult to separate the policy theorist from the man—which makes this book partly, and inescapably, a memoir. The six chapters are chronological and, to a large extent, thematic. In each new phase of Marburger's career, the tasks he faced brought him into contact with yet another dimension of science policy, in addition to the ones with which he was already familiar.

Chapter 1 is about Marburger's role as chairman of the Shoreham Nuclear Power Plant Commission. "It was the Shoreham Commission that opened my eyes to the world of policy," he writes. The Shoreham Nuclear Power Plant was a $3.4 billion facility intended to produce a useful commodity—electricity—but was strongly opposed by anti-nuclear power activists. In 1983, New York State governor Mario Cuomo appointed a "fact-finding panel" to study the economics, safety, and policy

---

[2]John H. Marburger III, "On Management II," September 23, 1982. John H. Marburger III Collection, Box 7, Stony Brook University Special Collections and University Archives.

INTRODUCTION   5

issues of the plant. Marburger had arrived at Stony Brook just three years earlier, but Cuomo, respecting Marburger's already evident ability to handle controversies, appointed him the panel's chairman. The panel included a broad range of Long Islanders, from antinuclear activists to scientists, whom Marburger (and Cuomo) knew would never agree on this issue. "Not since Captain Bligh has anyone commanded a crew trying to sail in so many different directions," Cuomo wrote Marburger. Asking himself what good might come of a committee sure to be deadlocked, Marburger decided that the most productive step was, in the final surprising report, not to attempt to adjudicate or even make recommendations, but to describe a larger vision of the events in which even the most radical partisans could recognize themselves. That effectively provided legitimacy to any decision Cuomo eventually made (which was ultimately to decommission the plant). This illustrates another distinctive element in Marburger's approach: it is important, especially when dealing with controversial policies, to embed policy decisions in a story respected by all participants, that is, to provide a *recognizable narrative* of the background events.

Chapter 2 concerns Marburger's role as chairman of the board of trustees of the Universities Research Association (URA), the group charged with managing the Superconducting Super Collider (SSC). He became chairman of the URA Council of Presidents in 1987 and then, when URA created a board of trustees, its chairman in 1988. The SSC project ran into trouble, partly due to the impact of new government procurement requirements on how large scientific projects were proposed, approved, and constructed, and it was ultimately terminated by Congress in 1993. This experience impressed him with the inevitability of public oversight of large scientific projects and with the need for planners of such projects to establish and make public a cost and schedule tracking system that would model the project's progress and all of its expenditures. Many science administrators regarded this as a potential obstacle when constructing daring, state-of-the-art scientific facilities— which are deliberately risky and consciously push the technological envelope, and therefore inevitably encounter unforeseen difficulties. But Marburger regarded providing such a public tracking system as a necessity for large scientific projects in today's environment, in a policy principle that he called the Principle of Assurance.

Chapter 3 is about Marburger's activities as director of Brookhaven National Laboratory. Marburger became Brookhaven's director thanks to an episode that began in 1997, when a small leak of slightly radioactive water from the spent-fuel pool of a reactor at the laboratory led to a media and political uproar, leaving in its wake the firing of Brookhaven's contractor by the Department of Energy, the resignation of Brookhaven's director, the termination of the reactor, and calls by activists to close the lab. Marburger was eventually chosen as the lab's new director and was charged with handling this controversy. The selections in the chapter are instructive for how Marburger spent his executive time. They contain, for instance, Marburger's response to a letter from a middle school student that others might have dismissed—but Marburger found it worthwhile to compose a lengthy and thoughtful reply.

Meanwhile, he faced other controversies at Brookhaven, including the widely publicized suggestion that its newest facility, the Relativistic Heavy Ion Collider, would create a black hole that would destroy the universe. Included is one of Marburger's imaginative and compelling "thought experiments" to convey the meaning of a small amount of uncertainty connected with the risk, using the Loch Ness monster as a model. One of his strategies while dealing with Brookhaven's controversies was to proceed as he did during the Shoreham Commission: to repeatedly paint a big and convincing narrative of the controversies in which all sides could recognize themselves. Marburger carefully composed these narratives, so that they were clear and informative to non-scientists and yet not patronizing. This was no small feat. It is often dreary and uninteresting for outsiders, in controversies marked by extreme declarations and pageantry aimed at the media, to listen to technical details, fathom the often mind-numbing units of measure and decipher their meaning, and appreciate references to background literature; far more appealing are images of Frankensteins, mushroom clouds, and apocalypse. But Marburger's conscientious narratives allowed those who followed them to appreciate the significance not of the amplitude of specific voices in the debate but of where the individuals who

INTRODUCTION   7

were speaking were standing. He made it possible to notice the acoustics of the controversy, so to speak.

Marburger also realized that it was important to preserve his own credibility. Once he happened to be at the lab in jeans on a weekend when a media firestorm broke. Without his usual suit jacket and tie, he was unwilling to appear in a way that might be read as informal and unprofessional, and while he agreed to speak to reporters, he declined the invitation to stand in front of video cameras. As Marburger wrote later, he learned to keep his image under his own control, noting that the most credible and authoritative public image for an administrator involved in a controversy was usually "as bland a one as possible."

In 2001, while he was director of the Brookhaven National Laboratory, Marburger took the most controversial job of his career: science advisor to President George W. Bush, in which post he was also director of the White House Office of Science and Technology Policy. Here the challenge—the burden of representation—was still greater and more complex. Marburger was in the position of having to represent the scientific community and its findings to the administration, and he had to represent the Bush administration, which was free to accept or ignore these findings, to the scientific community. Many scientists found these two roles orthogonal, even antithetical. Marburger decided to go to the White House as a scientist, not as an advocate. He did not waste energy on high-profile, ultimately political controversies that were not in his power to influence—Bush's stem cell research decision, for instance, or the alleged science offered to justify the U.S. invasion of Iraq. Instead, Marburger took on important issues in his power. Noticing the extremely significant but usually overlooked fact (pointed out to him by science policy expert Daniel Sarewitz) that, for decades, the federal science budget has been highly stable as a fraction of the gross national product regardless of the administration, Marburger realized that he had a real chance to make a long-term impact on U.S. science funding not by trying to increase the science budget but by increasing how efficiently this money flows through the dozens of uncoordinated agencies.

Chapter 4 is about Marburger's engagement during President Bush's first term, in which he was concerned with defining the roles of the presidential science advisor and Office of Science and Technology Policy director, with addressing issues caused by the impact of homeland security regulations on the science and technology workforce, and with handling severe criticism directed at the Bush administration for its handling of science. In 2004, an election year, these accusations reached their harshest moment in a document called "Restoring Scientific Integrity in Policy Making," signed by sixty-two eminent scientists, that charged the Bush administration with "suppressing, distorting or manipulating the work done by scientists at federal agencies."[3] After Marburger responded to the document, unprecedented criticism was directed at him in a sometimes vitriolic and personal way, including most notably the charge by the Harvard psychologist Howard Gardner that Marburger had become a "prostitute" by virtue of working for the Bush administration. Yet Marburger (who took a salary cut to go to Washington) remained unflappable and barely registered the criticism publicly or even privately.

Chapter 5 is about Marburger's activities during Bush's second term. During this period Marburger highlighted the importance not only of prioritizing scientific projects in federal science policy but also of prioritizing the particular areas within science that were appropriate for the federal government to support. "The word 'prioritize' sends shivers down the backs of most science advocates," Marburger noted, but as a former university president and laboratory director, he realized it was inevitable, was quite comfortable with the idea, and knew how to create a legitimate process to carry it out. The results of his prioritizations of certain general areas of science for federal funding are embedded in the American Competitiveness Initiative, crafted by Marburger and announced by President Bush in his State of the Union address in 2006. Another key policy initiative that Marburger promoted during this time

---

[3]"Scientific Integrity in Policymaking: An Investigation into the Bush Administration's Misuse of Science" (Union of Concerned Scientists, 2004), 5, www.ucsusa.org /assets/documents/scientific_integrity/rsi_final_fullreport_1.pdf. The accompanying statement, titled "2004 Scientist Statement on Restoring Scientific Integrity in Policy Making" (www.ucsusa.org/scientific_integrity/abuses_of_science/scientists-sign-on -statement.html), was later signed by more than 1,500 scientists.

INTRODUCTION   9

was the "science of science policy," an attempt to generate more measures and other tools to help policy makers arrive at effective policy
decisions.

In November 2006 Marburger was diagnosed with non-Hodgkin's
lymphoma, and the following month he began chemotherapy. Nevertheless, he continued to work uninterrupted, even from his hospital room in
New York City while undergoing treatment. In July 2007, a few weeks
after receiving a strong dose of a radionuclide therapy, he set off the
radiation detectors while entering the White House and was startled by
guards suddenly appearing in front of him, half-crouched, to block his
way. Marburger persevered with his work on science policy after leaving the White House in 2009, at the end of President Bush's second
term, returning to Stony Brook University to become vice president for
research. Marburger also found time to write a book on quantum mechanics, whose scientific and philosophical intricacies were one of the
passions of his life: *Constructing Reality: Quantum Theory and Particle
Physics* (2011).

I first met Marburger shortly after I joined the Stony Brook faculty in
1987. In the 1990s, I began working on a book about the early history of
Brookhaven National Laboratory. After Marburger stepped down as the
university president in 1994, we had breakfast or lunch at regular inter-
vals to discuss the lab's history. These meals were always at diners; he
explained that it was important to him to be accessible to the community, and eating in diners maximized the opportunity to encounter
people. I was often astonished at the kind of advice he would give me,
including what kind of tie and jacket to wear when giving a talk in order
to convey authority to a particular audience. He and I ended up coteach-
ing a National Science Foundation–sponsored course at the university,
titled Social Dimensions of Science, as part of Project WISE (Women
in Science and Engineering). But I first appreciated Marburger's leadership instincts, and his profound grasp of science in society, during the
controversy of 1997 at Brookhaven. I would show up at the diner expressing incredulity at some recent episode at the lab, and Marburger
would then laugh at my innocence and would carefully explain to me
how all the actors involved—including the activists, lab management,

and politicians—were constrained by forces acting on them, why they had very little room to act otherwise, and why whatever had just transpired was all but inevitable. I was astounded; imagine watching a play whose events seem utterly chaotic, and then hearing a critic explain persuasively the rationale behind each and every onstage action. During one such meal, Marburger described his vision to me as "Spinozist," alluding to the Dutch philosopher Baruch Spinoza's view that the world is like a vast ecosystem, all of whose elements have grown into the niches that they occupy, and they continue to unfold and act on each other the only way they can. Public controversies are like that, Marburger said. To intervene successfully, a leader has to understand the ecosystem very well, and to move very slowly and carefully.

During Marburger's years as presidential science advisor I met with him infrequently, but we resumed our diner meals after he returned to Stony Brook. Shortly after finishing *Constructing Reality*, he began to write a book about science policy based on his own experiences. In June 2011, after writing two chapters, he asked me to finish the book, should he be unable to himself, by editing material from his talks and writings. He gave me a rough outline. A few weeks later, in July, he passed away. This volume completes what Marburger started. The first two chapters are the ones he deemed finished; subsequent chapters consist of his most important talks and other reflections on science policy. I have written introductions to the chapters and the selections, and my contributions are in a different font to clearly distinguish them from Marburger's.

Science policy, in Marburger's eyes, involves far more than making wise decisions about the appropriate size of a science budget and how best to apportion it—"How much should go to the NSF and how much to the NIH?" "How much should go to nuclear physics and how much to malaria research?" It also requires making sure that decisions are successfully implemented. Marburger's words and actions illustrate a range of strategies for doing so: respecting the distinction between advocacy and advice, embedding science policy decisions in a recognizable narrative, earning the community's trust in areas where scientific facilities are located, heeding the Principle of Assurance, assuring the

INTRODUCTION   11

credibility of the policy maker, being unafraid to prioritize, and seeking to develop measures and other tools for making good decisions about
policies. Taken together, these writings constitute a sweeping and original approach to science policy, whose elements will be indispensable for
meeting twenty-first-century challenges.

The final chapter contains two essays Marburger published while at
Stony Brook. One, an essay he wrote for a book he coedited called *The
Science of Science Policy: A Handbook,* is about the importance of developing measures and other tools for science policy. The second concerns
the peculiar kind of authority that science has in politics. Invoking sociologist Max Weber's three kinds of authority, Marburger points out
that the authority of science in government circles has to be considered
"charismatic," or based on its perception as "endowed with supernatural, superhuman, or at least specifically exceptional powers or qualities." This will sound absurd to scientists and many others, Marburger
writes, for "it is precisely because the operation of science does not require charismatic authorities that we should trust it to guide our actions,"
and it is normal to regard acting against the authority of science as "a
mild form of insanity." Still, he continues, unless the authority of science "is enforced through legal bureaucratic machinery," the cold truth
is that "science is a social phenomenon with no intrinsic authoritative
force." Therefore, science practitioners and policy makers must continually guard their credibility to be effective, and "science must continually justify itself, explain itself, and proselytize through its charismatic practitioners to gain influence on social events." This book is an
effective guide to how it's done.

1

The Shoreham Commission

John Marburger assumed the presidency of Stony Brook University in 1980. An article about him that appeared in *Newsday* a few years after his arrival, titled "Tact Is a Science for Marburger," described him as having "graying dark hair and a movie star's blue eyes and square jaw" but observed that a key to his success was that "he is not afraid to be boring." The article said most faculty members appreciated that Marburger was not as confrontational as the previous president but reported that a few professors did not like the fact that he appeared "cold and unemotional" and overly formal, never without jacket and tie.[1]

At Stony Brook, Marburger faced yet more volatile controversies than he had at his previous position as dean of the College of Letters, Arts and Sciences at the University of Southern California. One began when an African studies professor equated Zionism with racism, inciting calls from the conservative Jewish community in the New York metropolitan area for the professor's firing and for Jewish students to arm themselves in self-defense when on campus; this provoked angry replies from the African-American community. Eventually, Marburger managed to defuse this crisis with a miniature version of the aggressive listening that would mark his later style, which left all sides feeling understood. He denounced the equating of Zionism and racism, but also, after a formal investigation, refused to fire the faculty member, though advising him to include more balance about the issue in his lectures and assignments. As the *Newsday* article reported, "Mar-

---

[1]Michael D'Antonio, "Tact Is a Science for Marburger," *Newsday*, November 11, 1983, 6.

THE SHOREHAM COMMISSION   13

burger crafted public statements that reflected both the concerns of Jewish leaders outside the university and academics within. He brought the faculty Senate into the debate and met repeatedly with Jewish leaders, eventually caucusing with from 30 groups [sic] in a single meeting. The result was a declaration that both satisfied Jewish groups and reassured professors who were concerned about protecting [the faculty member's] academic freedom."[2]

Marburger's skill at handling conflicts did not go unnoticed, including by New York State's new governor, Mario Cuomo. In 1982, Cuomo was elected governor of New York and took office on January 1, 1983. One of the first crises involved the Shoreham Nuclear Power Plant. The Long Island Lighting Company (LILCO) was in the process of building this facility near the town of Shoreham, about a dozen miles farther east on Long Island than Stony Brook. Hearings on the construction permit began in 1970, the permit was approved in 1973, and the reactor pressure vessel was set in place in December 1975, amid growing opposition. Demonstrations against the plant commenced the following year and gained momentum after the 1979 Three Mile Island accident. When Cuomo took over in 1983, the controversy was still continuing without a resolution in sight. Cuomo created a fact-finding commission and appointed Marburger to head it.

Heading the Shoreham Commission was Marburger's first exposure to a major political crisis involving a scientific and technological dimension. It thrust him, Marburger writes in this chapter, into "the universe of public action" and "opened my eyes to the world of policy." A common reaction for a manager facing a crisis is to figure out a solution, discover who opposes that solution, and then do battle with them. That approach was not Marburger's. His attitude was to say, "Isn't it amazing that reasonable people have such different points of view on this issue! How is that possible?" He would then set out to address that question, which often opened up new, unanticipated opportunities for a solution.

Marburger's experience as head of the Shoreham Commission would shape the way he handled later crises. It would also give him a reputation as an impartial and calm leader. The following is a chapter, printed in its entirety, that Marburger completed for this book. It consists of his comments on the Shoreham episode, interspersed with three inserts that he had selected: a draft he wrote in August 1983 of an introduction to the Shoreham Report,

---

[2]Ibid, 23. Later, however, the professor involved was denied tenure at Stony Brook.

14   SCIENCE POLICY UP CLOSE

which was partly included in the final report of the Shoreham Commission;
a memo to himself he wrote in December 1983 after the report was released;
and an article about him that appeared about the same time in the *New York
Times*.[3]

—RPC

In the Beginning, There Was Shoreham

Telephone conversation, President's Office, Stony Brook University,
March 15, 1983

"Jack, this is Hank Dullea" [director of state operations for newly
elected New York governor Mario Cuomo]. "You know that LILCO has
asked the state to substitute for Suffolk County in the Shoreham nuclear plant's off-site emergency plan. The governor wants to form a fact-
finding commission, and he wants you to chair it."

"Hank, that's very flattering, but this is a hugely controversial issue
here on Long Island, a real lose-lose situation. My involvement in something like this is bound to have an impact on Stony Brook [University]."

Pause.

"Jack, the governor can have an impact on the university, too."

Pause.

"Hank, I think you're speaking my language."

Thus are scientists called to public roles.

I accepted the task, worked with the commission from May to December 1983, and produced a report that, in its complex tangle of answers to
questions, consensus points, huge digressive minority reports, staff re-

---

[3]The title page of the final report, a government publication in the public domain,
reads as follows: "Report of the New York State Fact Finding Panel on the Shoreham
Nuclear Power Facility, Honorable Mario M. Cuomo, Governor; Dr. John H. Marburger, III,
Chairman, Stony Brook, New York, December 1983." John H. Marburger III Collec-
tion, Box 11, Stony Brook University Special Collections and University Archives, Stony
Brook University, NY.

THE SHOREHAM COMMISSION   15

ports, lists, and tables, invites comparison with Goethe's *Faust*. Like *Faust*,
the report had drama and a thread of philosophical reflection, the latter resulting from a habit of mine to generalize immediately from experience to fit real-time action into a broader picture to guide my own
next steps. The report is available in the archives of the Stony Brook
University Library, and I will not attempt to summarize it (my draft introduction follows). But it was the Shoreham Commission that opened
my eyes to the world of policy, and I reproduce some documents below—
interviews, excerpts from reports I wrote, and memos to myself—that
will shed some light on this, my introduction to a universe of public
action entirely new to me.

Dullea's call came early in my term as Stony Brook's president, but it
was not the first such. Suffolk County's executive Peter Cohalan had
twisted my arm in August 1980 to head a committee on the future of
county finances. Just a few months before, as an active scientist and
dean of Letters, Arts and Sciences at the University of Southern California, my head was full of physics. I knew nothing of public finance,
but I saw Cohalan's offer as a chance to learn the new territory that
would turn out to be my base for the rest of my active life. I managed
the meetings as I had attempted to manage faculty meetings—with
fairness and patience, counting on expert advice but knowing I would
be interpreting the history of the event. My good fortune in this case
was having an experienced guide, a senior colleague on the committee
with deep knowledge of the county and its politics and economics.[4]

Luck in counsel helped again when Governor Hugh Carey, Mario
Cuomo's predecessor, prevailed upon me in December to chair a "sun-
set commission" on New York's State Energy Office. I had excellent
staff support from the relevant state agencies and from the highly com-
petent director of the office itself, whose continuation the commission

---

[4]My mentor on the Suffolk County finance panel was Lee E. Koppelman, Suffolk
County's widely respected regional planning director since 1960, a colorful, quotable
visionary who dominated planning discussions on Long Island for more than forty
years.

unanimously recommended.[5] These early experiences brought me into direct contact with the issues and the politics into which I would plunge in just a few months with the Shoreham Commission. Who could guess then that I would be dealing with much the same community fourteen years later, exercised by safety issues once again at yet another nuclear reactor at Brookhaven National Laboratory?

Draft Introduction to the Shoreham Report

[August 1983, written for the commission members,
partly incorporated in the report]

The Shoreham Nuclear Power facility is the focus of an extraordinary and debilitating controversy in the two eastern counties of Long Island. Of the hundreds of questions raised about it during the fifteen years since it was first conceived by the Long Island Lighting Company, there are three basic ones that now demand resolution.

1. Should the Shoreham facility, now essentially complete, be allowed to operate?
2. Whether it operates or not, who should pay for it?
3. How will these questions be answered?

That such questions should be voiced at all is extraordinary. There is no precedent for the abandonment of a $3.4 billion facility that produces a useful commodity. There is usually no doubt in the utility industry about who should pay for what. And the decision-making structure for deciding these issues has seemed until now very well defined.

The reason these questions are being taken seriously is that those asking them know that they will suffer financially as a result

---

[5]James LaRocca, the first New York State commissioner of energy (1977–1982), subsequently served in important capacities in state government and Long Island business and academic affairs.

of the usual decision-making process, and many of them believe
in addition that their health and that of their descendants will be
endangered if the facility operates. They see the Shoreham plant as
having been thrust upon them unnecessarily by a profit-seeking
entity, and they are attempting to deflect its consequences through
the power that they believe they should have in a democratic
society. They believe that their elected leaders should heed their
concerns and alter the normal processes, if necessary, to abort the
certain financial impact and the possible health impact to which
those processes appear to be leading.

The governor's fact-finding panel was formed to disentangle and
clarify the issues contributing to the Shoreham controversy, and
thus to assist the governor in choosing a course of action for the
state of New York. In performing this task, the panel must attempt
to distinguish between what the various parties assert and what is
actually the case. That is, unfortunately, not an easy task. It is complicated, first of all, by the universal tendency of those who seek an
end to advance all possible arguments toward that end, regardless
of the quality of the argument, and second, by the fact that, while
most assertions are about what will happen in the future, the future
is very difficult to predict.

It was certainly the difficulty of predicting the future that led
LILCO to embark upon and then pursue the course that will lead to
the highest electricity rates in the country for its consumers. No
one anticipated the oil crises of the '70s and the rapid global
realignments of industrial activity that brought regional growth to a
halt while the power plant that was to serve it was under construction. And no one predicted either the fact or the consequences of
the accident at the Three Mile Island nuclear facility, an incident
that contributed enormously to the direct costs of the Shoreham
plant and to the uneasiness of its neighbors.

The difficulty of predicting the future weakens the usefulness
of the most common criterion for public action: choose the course
that brings the most benefit to the most people. Utility planning
extends over such long periods of time that benefit assessments are

unreliable in the extreme. All that can be done is to assume that the future will be much like the present and the immediate past and make the best guesses that one can.

In the natural sciences, prediction is more reliable and better defined than in economics, but it is also a more technical concept. Scientific prediction is nearly always statistical prediction, whose accuracy depends upon the weight of experience and the completeness of the predictive model. And yet there are great laws of science, exceptions to which have never been observed. The mix of certainty and uncertainty in science is a source of confusion to a public whose view of science is idealized. Much of the debate over the safety of nuclear power plants in general, and Shoreham in particular, centers upon the significance of a wide variety of statistical predictions.

The broad issues of economics and safety affect each other to some extent, but can be analyzed separately. The time of the panel was divided among these topics, and our report will also treat them separately.

The panel met twelve times and conducted four public hearings in Albany and across Long Island. As the hearings moved from west to east, nearer the proposed facility, they became more heated and contentious, challenging my ability to keep order. At one point I seized a microphone from the hands of a panelist; at another, in Riverhead, a few members of opposing factions rose from their seats, shouting and shoving at each other. My aide, Patricia Roth, a woman of impressive determination, defused the situation by striding down the aisle between them, elbows out.

The meetings were no less passionate than the hearings and were plagued by leaks that spun toward the opposition as the report neared completion. County executive Peter Cohalan was running for reelection on November 8, stressing his opposition position, and I shut the committee down to ride out the elections, scheduling the release of the report for mid-December. On November 21 news of a massive leak appeared on the front page of the *New York Times,* complete with my picture above the fold, slanted toward the opposition view. When the report appeared, the press was puzzled by the absence of recommendations.

Remarks on the Reception of the Shoreham Report

[December 20, 1983, JHM memo to file]

That the governor's fact-finding panel on Shoreham was divided is the least surprising thing that could be said about it, but it was the feature that has drawn most attention in the press. It was certainly divided from the beginning, given the strong differences among important actors: LILCO, Suffolk County, the Nuclear Regulatory Commission, local citizens groups, business associations. There were even significant differences among "expert" witnesses and consultants. Each of these had been studying Shoreham for years, and it was unlikely that a few months on a special committee was going to change anyone's mind.

Governor Cuomo understood the inevitability of this disagreement even as he conceived the panel. In every statement he issued as it was formed, in the charge to the group, in instructions to his staff, he cautioned against efforts to arrive at conclusions. He said he wanted facts and ideas so that he could make the decisions himself. As chairman of the commission, I echoed that approach, which at least one reporter thought so remarkable last June that his story carried the headline "Shoreham Panel to Make No Recommendations." That principle appeared again in the governor's formal charge to the committee, included in the introduction to the report, and is reinforced by my transmittal letter to the governor and again by the press release that came out with the report. My own remarks at the well-attended December 14 press conference were almost exclusively devoted to the idea. The governor did not select the panel with his eyes closed. He knew there would be disagreements. He wanted input from all sides. He did not want, did not ask for, a grand conclusion to slice through the Shoreham knot.

The panel, of course, was full of intelligent and incisive people: academics, scientists, economists, public advocates, businessmen, and the most brilliant and articulate among the lay opposition. But in the final analysis there are only two positions on Shoreham and, indeed, on nuclear power in general. There is the Establishment Position and the Opposition Position. Which one you pick depends on whom you trust. The Establishment view is advanced by the overwhelming majority of scientists and engineers and by nearly all technical people who know anything about nuclear power. It holds that the risks posed by nuclear reactors are negligible compared with many other risks that society routinely accepts and that nuclear power is a clean, safe, attractive alternative to generating electricity by burning coal or oil. The Establishment view has its skeptical adherents, who see problems in the way nuclear technology is managed and regulated, but they believe those problems are solvable and generally of more consequence for economics than for public health.

The Opposition view is that nuclear power has been forced on the American public by a heedless governmental/industrial complex of such influence that it guides the very course of the scientific enterprise itself. Only a few objective scientists, willing to risk loss of jobs and research support, are laboring to expose the fallacious reasoning of the Establishment scientists who support the aims of this complex. Some adherents to this view believe that nuclear power could be rendered safe someday but that known dangerous side effects are being ignored by those currently responsible for safety regulations. Other characteristics of this view are described in the admirable book *Risk and Culture* by Mary Douglas and Aaron Wildavsky (1982).

As for the Shoreham plant, the Establishment sees the Opposition as not having demonstrated its dangers. The Opposition sees the Establishment as not having demonstrated its safety. Both are right, because safety and danger involve the concept of risk, and

THE SHOREHAM COMMISSION

risk involves probabilities. The Establishment view of the probability of a serious accident is a realist one: compared with other risks, it is negligible. The Opposition view is that the consequences of even an unlikely accident could be so severe that the risk is not worth it.

Could an accident at a nuclear power plant be as severe as the Opposition suggests? Could a Shoreham accident ever be so bad as to require a twenty-mile evacuation planning zone? It depends on whom you trust. This is a technical question, whose answer requires input from many disciplines, including medicine, meteorology, physics, chemistry, and mechanical engineering. Each discipline has its own Establishment and its own Opposition. In the world of science, however, those camps have different significance. Science is a communal enterprise. When we speak of scientific knowledge, we do not mean the opinion of a single scientist; we mean the body of material mutually acknowledged by the scientific community. The technical answer to the technical question of safety is the Establishment answer.

Aren't scientific theories sometimes found to be wrong? And isn't it true that the right theory is usually anticipated by a scientist whose views were thought to be outside the mainstream? Yes, but in the natural sciences major changes in theoretical point of view are extremely rare, and when they occur, the valuable practical rules are found to be just as useful, but embedded in the new structure in a different way. It is significant that the consultants retained by Suffolk County to advise on risks and off-site emergency planning used generally the same methodology and arrived at generally the same conclusions as LILCO consultants. Press accounts emphasize differences. What is remarkable are the similarities. The differences are in degree, not in kind.

The Shoreham panel spent most of its time studying and discussing the safety and off-site planning issues because those are the ones holding up licensing of the plant. The Suffolk County executive has made it clear that the county's decision to refuse to cooperate in off-site emergency planning is based on its belief that no plan could guarantee the safety of all those whose health may be affected by a severe accident. Indeed, no plan could ever do that, even if the accident were fire or storm. The county consultants presented a conservative version of the Establishment view of the significance of risk. The county decision adopts the even more conservative Opposition view. The consultants did not say an emergency plan is impossible; Suffolk County did. Indeed, there is no technical evidence that an adequate emergency plan (an Establishment notion) is not possible for the Shoreham plant. On the other hand, the notion that such plans cannot guarantee everyone's safety (an Opposition requirement) is applicable to all emergency plans everywhere, which seems to me to be a weakness of the county's position.

These issues are laid out in the Shoreham report. It may not be an easy report to read, but it is a gold mine of information and summaries of important views about Shoreham and about nuclear power. It includes both Opposition and Establishment views in easily recognizable forms. The staff reports are extremely important and are not simply the product of bureaucrats working in isolation from the panel. The authors attended nearly all the panel meetings and attempted to reflect the priorities and orientations of panel members.

What is the governor to do with this prickly report? I think it is close to what he asked for, whether or not he expected it to look as it does. The emphasis on the safety issues is appropriate to the most difficult of the decisions he faces, which is how or whether the state will address the licensing impasse created by Suffolk County. The issues there are clear enough and spelled out in our report. Recent statements by the governor indicate that he has already begun to formulate a position based on his continuing insistence on federal involvement in the solution. There is much sense in this, because the federal government bears considerable responsibility for creating the situation in the first place.

Regarding the economics issues, which most Establishment observers see as by far the most important, there was never any

THE SHOREHAM COMMISSION   23

doubt that the state would have to be involved in rate shock
mitigation. The state already had arranged for cheaper electricity
for Brookhaven National Laboratory in 1981 and participates
heavily in the rate-setting process through its Public Service
Commission. The most important question about economics is
whether the plant has value as a source of electricity. The answer
is yes, almost certainly. The value is probably not as great as the
cost of the plant, but now that the plant is built, it makes economic sense to operate it. Other decisions regarding specific
strategies for rate mitigation must await resolution of the licensing impasse. The various scenarios depend sensitively on when
and under what conditions the plant begins to operate, and how
healthy LILCO is when it starts. If the licensing impasse remains
unresolved, the rate scenario will look completely different. The
governor will need expert advice as the regulatory machinery
cranks out Shoreham's fate. He has already begun to seek it, and I
am sure the analytical apparatus prepared by staff for our panel
will continue to be useful.

The Shoreham saga does not lend itself to generalizations. It
poses a difficult task for journalists who have the dual objective of
informing and titillating their readers. One gets in the way of the
other. But beyond the obvious dissension, there are some clear
choices set forth in the Shoreham report. The report is not simple,
but then neither is the governor for whom it was written.

These "remarks" are typical of essays I wrote to myself throughout my
life to clarify my thoughts. I rarely completed them, and the final paragraph in the original of this one is written out by hand on the last page.
As I gained more confidence in my public role, I shared my thinking
with a wider audience, but already at USC I had learned how public ut-
terances mutate under media scrutiny, and my instinct was to keep my
image under my control. The practice contributed to a bland public image, but my official responsibilities obliged me to attempt to influence
others, and human behavior ever resists an inconsistent leadership.

How the Shoreham Saga Ended

Remarkably, the Nuclear Regulatory Commission granted the Long Island Lighting Company a license authorizing fuel loading and cold criticality testing on December 7, 1984, followed by a low-power license in July 1985 and a full operating license on April 21, 1989. By that time, however, the political wheel had turned. The state of New York intervened consistently in the licensing process to oppose the operation of the facility, and the governor prevailed in hardball negotiations with LILCO's equally brilliant and tough-minded chairman, William Catacosinos. Here are the facts from the April 1990 decommissioning report prepared by the Long Island Power Authority (LIPA):

> On February 28, 1989, the State and LILCO entered into a Settlement Agreement under which LILCO agreed not to operate Shoreham and to transfer the plant to LIPA [in exchange for $1]. Thereafter, on April 14, 1989, LILCO and LIPA entered into an Asset Transfer Agreement under which LILCO reiterated its agreement never to operate Shoreham and to transfer the plant to LIPA. . . . The Settlement became effective on June 28, 1989 when LILCO's shareholders voted to approve it. . . . As required by the Settlement, LILCO has never used its full power license to operate Shoreham. In fact, LILCO commenced defueling the plant shortly after LILCO's shareholders voted to approve the Settlement, and completed that task on August 9, 1989. Shoreham has remained in a defueled state since that time.[6]

Why the LILCO shareholders agreed to give up what had grown into a $5.2 billion asset despite having the legal authority to operate may be inferred from the objection of the regional business community through its Long Island Association. LIA chairman John V. N. Klein, Cohalan's immediate predecessor as county executive, declared that the plan was

---

[6]Long Island Power Authority, "Shoreham Nuclear Power Station Decommissioning Report, April 1990, unpublished manuscript, p. 5. John H. Marburger III Collection, Box 11, Stony Brook University Special Collections and University Archives, Stony Brook University, NY.

THE SHOREHAM COMMISSION   25

too costly to ratepayers; unfairly apportioned the burdens of the settle-
ment; failed to take account of future contingencies; did not provide for
Long Island's electricity needs; and had been insufficiently debated and
scrutinized.

In his determination to prevent Shoreham from operating, the governor had to overcome the enormous leverage the NRC had created for
LILCO by granting it licenses. Negotiations between the governor and
LILCO, described to me by its chairman and others, were confrontational, blunt, and bitter. Both strong men achieved their objectives with
the assistance of government machinery set up at state and federal levels to foster contrary outcomes. LILCO avoided bankruptcy, and Shoreham died. The rate payers, not the shareholders, are paying for the abandoned plant. By 2009, according to *Newsday*, "$16 of every $100 paid in
Long Islanders' LIPA bills goes to debt service tied directly to Shoreham."[7]
NRC's effort to enforce the Establishment position, and the New York
Public Service Commission's effort to spare rate payers by denying significant rate increases until the plant was "used and useful," both added
substantially to the very high electricity rates on Long Island today.

The public had spoken through its democratic processes: the political
optics of the Opposition view were simply too powerful for the relatively
bloodless financial perspective of the Establishment position, however
fair or rational. The governor had to know he was in a politically awkward
position from the instant LILCO approached him for relief. It did not take
a Mario Cuomo to foresee that every elected official at every level of government who supported Shoreham would lose their seat in the next election. But the political, financial, economic, and safety issues had been
swirling about the facility for fifteen years, and it was not clear how
things would turn out. Even the March 1979 Three Mile Island nuclear
accident had not weakened LILCO management's confidence that the
plant would eventually operate. The Shoreham Commission had a moderating effect on the pace of events that played into Mario Cuomo's main
political strengths. In the end, Establishment and Opposition both cited

---

[7]Mark Harrington, "Saga behind the Shoreham Nuclear Plant Retold," *Newsday*,
June 13, 2009 (http://archive.today/jEdkT).

the Shoreham Commission report to make their various cases, and the governor found ample support in it for the position he had to know from the beginning was the only viable political option.

And then, of course, on April 28, 1986, there was the Chernobyl disaster. Little more than two months later, the New York State Legislature created LIPA, and the Shoreham saga wound to its inevitable conclusion. I would learn much more about Chernobyl two decades later to fulfill a presidential directive, but the core issues that plagued the Shoreham nuclear facility were much the same.

New York Times, **December 15, 1983**
Man in the News; A Mediator for Shoreham Panel:
John Harmen Marburger III

Michael Winerip, Stony Brook, L.I.

Last May, when Governor Cuomo named him to head the special commission on Shoreham, John H. Marburger III believed that by accumulating enough scientific data it would be possible to determine whether the nuclear power plant could safely open. But after conducting 17 commission meetings from Albany to Riverhead, the 42-year-old college president said, he came to feel that scientific answers were not enough.

"I originally felt there were ways of establishing failure rates and probabilities of various kinds of accidents," he said. "But it turns out it's much less quantitative than that. There's a lot of sociology involved."

Part of his original faith in solving the Shoreham equation with numbers may have been the physicist in Dr. Marburger. Until becoming president of the State University of New York at Stony Brook three years ago, he did research in applied laser physics, a field where formulas provide answers.

"An Incurable Optimist"

Part of it, too, may have been the carpenter in Dr. Marburger. He is capable of taking complicated plans and constructing a neatly finished product. He has, for example, built his own harpsichord.

And part of that faith just may have been the optimist in Dr. Marburger. "I'm an incurable optimist," he told seven different people who visited him or called him at his campus office with seven different problems one recent day.

Dr. Marburger has won praise for his work as commission chairman from advocates on both sides of the issue and from the Governor's office. Most people said they were surprised that he had been able to mold as much of a consensus as he had done.

The Shoreham panel is one of several hot spots he has found himself in lately. He also faces a budget crisis at the university that may force layoffs and has been kept busy mediating a campus dispute over whether a course linking Zionism and racism should be permitted.

An Effective Mediator

He is an effective mediator in such disputes, colleagues say, because he does not take his disagreements personally.

"Anger is not a useful reaction except symbolically, in a public-relations sense," Dr. Marburger said.

As chairman of the Governor's commission on Shoreham, he brought patience, dignity and thoroughness to the job, fellow commissioners say. Both Leon Campo, a longtime anti- Shoreham activist, and Dr. Herbert J. Kouts, a longtime Shoreham supporter and chairman of the department of nuclear energy at the Brookhaven National Laboratory on Long Island, describe the chairman in identically favorable terms.

"And if you can get both Campo and me to agree on anything, that's something," said Dr. Kouts.

For Dr. Marburger, the commission was an experience that profoundly changed the way he thinks of the relationship between science and the public. Before working on the panel, he said, he had never been to a public hearing.

"Suddenly I saw this in a nonscientific light," he said. "It didn't matter so much if the facts on evacuating the plant area were right or wrong. If it governs the way people behave, then the factual becomes almost irrelevant."

Born on Staten Island

John Harmen Marburger, III was born on Staten Island on Feb. 8, 1941. He graduated from Princeton in 1962 and received a Ph.D. in [applied] physics from Stanford University. He moved to the University of Southern California, where he chaired the physics department and eventually became dean of the College of [Letters,] Arts and Sciences. Only in the last two years have his administrative duties kept him too busy for physics research.

His biggest disappointment at Stony Brook, he said, has been the restraints put on him by the state bureaucracy. While he has promised to stay at Stony Brook for 5 to 10 years, he said, his experiences with the state government have made him reluctant to take a state position again.

If he was criticized as chairman of the Shoreham panel, it was for listening too much. "He's generally given a chance for everyone to talk, no matter how outrageous, perhaps to the point of excess," said Alfred Kahn, a member of the Shoreham panel and Cornell University professor.

Attended All Meetings

Dr. Marburger was the only commissioner to attend all the meetings. "One of my objectives was to avoid having the anti-Shoreham people abandon ship," he said. "I didn't think I should shut them up."[8]

When he was named, groups opposed to Shoreham regarded him suspiciously because of his science background and his close ties to the Long Island business community, which generally supports opening the plant.

---

[8]"Not since Captain Bligh," wrote Governor Cuomo in his later thank-you note (January 15, 1984), "has anyone commanded a crew trying to sail in so many different directions. That you were able to chart a responsible course through such stormy rhetorical reefs redounds to your credit."

THE SHOREHAM COMMISSION   29

"I think it is true I would be expected to have that bias, so I don't deny that," he said. "It meant I had to work harder to convince people I was being fair."

Dr. Marburger said he had been frustrated that the commission had required so much work that it had taken time that could have been spent with his wife, Carol, his two sons, John, 13, and Alexander, 11, his routine Stony Brook work, his carpentry, his physics, his reading of philosophy and his piano playing.

He would have preferred, he said, to call fewer hearings. "The minute-by-minute experience of all these meetings was very difficult for me because of my view of how to proceed efficiently."

But a month after the commission began its work, he no longer felt that efficiency was the goal.

"It became apparent to me this commission was one of the rare opportunities for the public to feel they were being heard and taken seriously," he said. "There were democratic considerations."

Panel Members

Dr. David Axelrod, State Health Commissioner. Karen S. Burstein, president of the State Civil Service Commission. Leon J. Campo, executive assistant for finance of the East Meadow, L.I., public schools. William J. Dircks, executive director for nuclear operations of the Federal Nuclear Regulatory Commission. Marge Harrison, co-chairman of the Long Island Progressive Coalition. Alfred E. Kahn, economist and consultant. Dr. Herbert J. Kouts, chairman of the department of nuclear energy of the Brookhaven National Laboratory. Dr. Paul Alan Marks, president of the Memorial Sloan-Kettering Cancer Center. David McLoughlin, deputy associate director in charge of state and local programs for the Federal Emergency Management Agency. Dr. William J. Ronan, vice chairman of C.C.X. Inc. of Carle Place, L.I. David J. Willmott, publisher of the *Suffolk Life* weekly newspaper. Prof. Hugh Wilson of Adelphi University.

2

The Superconducting Super Collider
and the Collider Decade

In 1988, John Marburger became chairman of the board of trustees of Universities Research Association (URA), the group managing the Superconducting Super Collider (SSC). Before then, Marburger had been a member of Council of Presidents of its member universities, and was named chairman of the council in 1987. Most presidents did not go to the annual meetings, but Marburger did. As a young president of a young university near a national laboratory, he found URA an "efficient entry to Washington circles that were important to Stony Brook's long term research interests." By the time Marburger was named chairman of URA's board of trustees, the SSC was already running into scientific, economic, and political trouble, and it was canceled by the U.S. Congress in 1993.[1]

Until the SSC's termination, Congress tended to leave large scientific projects untouched. Before the final quarter of the twentieth century, in fact, science (physics in particular) had been largely protected from public scrutiny and government supervision. Scientists were both actors and judges of their own performances—a way of staging science set up by the government after the experience of the Manhattan Project to build the atomic bomb in World War II. But by the 1980s the Department of Energy (DOE) was exerting increased pressure for accountability on the large scien-

---

[1] See Michael Riordan, Lillian Hoddeson, and Adrienne W. Kolb, *Tunnel Visions: The Rise and Fall of the Superconducting Super Collider,* University of Chicago Press, forthcoming (note added by editor).

tific projects it funded.[2] The SSC's termination was a traumatic event for the physics community, reflecting a major shift in that community's relationship with both Congress and the executive branch.

Many accounts of the SSC saga attribute its demise to a combination of transitory factors, including poor management, rising cost estimates, the collapse of the Soviet Union and thus the end of the Cold War threat, and complaints by "small science" that the SSC's "big science" was consuming their budget, Congress's desire to cut spending, a new Democratic president (Clinton) who did not want to continue a project begun by two of his (Republican) predecessors, and unwarranted contract regulations imposed by the DOE in response to environmental lapses at nuclear weapons laboratories. Marburger, in contrast, tells a subtler story whose implications for science policy are more significant and far-reaching. His story involves changes in the attitude of the government toward large scientific projects that reach back to management reforms introduced by the administrations of presidents Johnson, Nixon, and Carter in the 1960s and 1970s. These reforms were unsuccessful, but when Japanese companies succeeded in imposing management reforms, increasing their competitiveness, U.S. manufacturers took notice and soon copied the reforms; this eventually led to a renewed effort by the U.S. executive branch to introduce contract reform measures in the 1980s. These reforms, which included increased demands for accountability, additional liability for contractors, and sanctions for infractions, were clumsily handled by the government and stoutly opposed by contractors, but the reform movement would culminate in the 1993 Government Performance and Results Act.

This, Marburger writes, is the climate that surrounded negotiation of the URA contract, in which the government insisted on including elements of these management reforms. Thanks in part to its price tag in the billions of dollars, the SSC project encountered this highly visible public accountability machinery—a first for accelerator builders. Marburger observes that many scientists, in the grip of what might be called "Manhattan Projectitis," felt blindsided and betrayed and condemned the application of contract reform

---

[2]See, e.g., Catherine Westfall, "Retooling for the Future: Launching the Advanced Light Source at Lawrence's Laboratory, 1980–1986," *Historical Studies in the Natural Sciences* 38 (2008), 569–609; Westfall, "A Tale of Two More Laboratories: Readying for Research at Fermilab and Jefferson Laboratory," *Historical Studies in the Natural Sciences* 32 (2001), 369–408 (note added by editor).

practices to the SSC as unnecessary bureaucratic interference, viewing URA's eventual decision to engage industrial partners to allay some of the pressure as cowardly capitulation. But Marburger argues that physicists could not have opted out of the bureaucratic procedures that were by then part of all large publicly funded projects; the momentum for contract reform was just too strong. Scientists now had to work in a theater whose imaginary "fourth wall," between actors and audience, had disappeared. The government and the public could now interact in significant ways with how science operated.

Marburger learned two big lessons about science policy from this saga. The first was an appreciation of the trajectory of the government's management reforms—that its impact on large scientific facilities and projects could not be reversed. The second lesson concerned exactly what these contract reforms were seeking: cost and schedule tracking systems to provide measures of program performance and impact. In practice, this would mean that each large project would need a simulation, or model, against which actual progress and expenditures could be compared. In the private sector, this is complicated but feasible; the baselines can always be shifted, and the model need not be openly released. In the public sector, however, this is high risk, because the model (a) implies baselines of quality, expertise, economy, and safety, with changes suggesting failure or incompetence; (b) must be approved by a public overseer; and (c) can be highly visible and the subject of congressional testimony. Indeed, once the SSC project got in trouble, and congressmen and executive branch members realized that URA had not yet implemented a functioning cost-schedule control system, it fed the perception of lack of public transparency, a large factor in the cancellation of the SSC.

Marburger knew that this state of affairs was messy, expensive, and inefficient but that science administrators had to embrace it. Here was another element in Marburger's approach: the realization that the disappearance of the fourth wall in scientific projects entailed the need to understand and adopt new parameters of accountability and transparency in what he called the Principle of Assurance. The length and complexity of this chapter—the mounting acronyms, unfortunately, are necessary!—reflects his feeling that the necessity of this lesson can be grasped only if he walks his presumably reluctant readers step by step through the entire story.

—RPC

THE SUPERCONDUCTING SUPER COLLIDER   33

The Demise of Isabelle

Driving east on the Long Island Expressway, Stony Brook University is
at exit 62, and Brookhaven National Laboratory (BNL) at exit 68. Established in 1947 as the Manhattan Project, and then morphed into the
U.S. Atomic Energy Commission (AEC), BNL was the prototype for today's Department of Energy (DOE) government-owned, contractor-
operated research laboratories and the model for CERN, its European
counterpart.[3] Its contractor for fifty years was Associated Universities,
Inc. (AUI), a consortium of nine distinguished private research universities created in 1946 by the New York State Board of Regents to "acquire,
plan, construct and operate laboratories and other facilities" serving the
research community and the federal government.[4] The first twenty-five
years of its history are chronicled in Robert P. Crease's *Making Physics:
A Biography of Brookhaven National Laboratory, 1946–1972* (1999).

BNL is famous among physicists, and its proximity to Stony Brook
strongly influenced my move to the university in 1980. Research at BNL
has a broad reach, but at the time I was fascinated by dramatic progress
in particle physics. Beautiful and intricate ideas had merged rapidly in
the prior decade to form a picture of matter called the Standard Model—
something like the chemical periodic table, but based on subnuclear
components called *quarks* and *leptons*. As Crease's title suggests, BNL
discoveries provided keys to this progress, enabled by ingenious particle accelerators and detectors developed at the lab, of which a machine
called Isabelle was to have been the next generation. It was competing
with a project at Fermilab near Chicago, the nation's largest accelerator laboratory, to be the first to use superconducting magnets,[5] and
with the already operating Super Proton Synchrotron, a powerful non-
superconducting machine at CERN. A third large device of a different

---

[3]CERN stands for Conseil Européen pour la Recherche Nucléaire, that is, European
Organization for Nuclear Research.

[4]Those nine research universities were Columbia, Cornell, Harvard, Johns Hop-
kins, MIT, Princeton, Rochester, University of Pennsylvania, and Yale.

[5]In a collider, particles are accelerated by magnets in opposite directions so that
when they meet head-on they collide with twice the energy of a "fixed target" collision.

kind was under construction at Stanford's Linear Accelerator Center. These are big machines. Isabelle was a ring about two and a half miles around, visible to airline passengers descending over Long Island toward New York's JFK Airport. Fermilab's Tevatron is nearly four miles around. Stanford's Linear Accelerator runs for two miles up to the edge of the San Andreas fault. In 1983 CERN broke ground for a seventeen-mile tunnel that would henceforth house the world's largest accelerators.

That nations should launch such huge projects to study realms remote from human needs is a curious but thoroughly human phenomenon. Pyramids and cathedrals come to mind, but these had their social purposes, and likewise (presumably) Stonehenge, Machu Picchu, and similar astrological complexes in Asia. Such expenditures in modern times usually go for national defense or civic works, such as dams or public transportation, but particle accelerators are for research that has been called "curiosity driven." In the half-century after World War II every large developed nation invested in "big science" projects whose relevance to other pressing social problems was not clear. The question of relevance comes up in congressional hearings and science policy discussions and confounds cost-benefit analyses. I wrestled with it for the three decades when my career placed me between those who believe and those who need to be persuaded that the enterprises are worth their costs. Although I am a believer, my purpose here is not to persuade but to recount my engagement with these magnificent endeavors. The story of the collider decade has multiple threads, and mine begins with Isabelle.

By 1981 Isabelle was in serious trouble for two related reasons. First, the pace of discovery had been faster than policy makers expected, and the experimental situation now demanded a more powerful, next-generation machine. Second, funding shortfalls and technical delays, not all unavoidable, had stretched the Isabelle construction schedule, making it unlikely that the completed machine would be finished in time to compete for significant discoveries (see Crease 2005).

Shortly after I arrived on Long Island, Stony Brook physicists, including Nobelist C. N. (Frank) Yang, visited me to express concern that not enough was being done on the political side to keep Isabelle fund-

THE SUPERCONDUCTING SUPER COLLIDER 35

ing in the federal budget. The economy was in recession from linger-
ing effects of the oil and energy crises of the 1970s, and the administration of newly elected President Reagan was proposing drastic cuts in
agency budgets to balance tax reductions and increases in defense
spending. DOE basic research funding was likely to be flat or declin-
ing, and priorities had to be established. Given the budget climate, I was
skeptical of our ability to have much impact, but the cause was important to Stony Brook, and I agreed to help. For me, however, the most
instructive episode of that campaign was a visit to Washington in September 1981.

Stony Brook and BNL are both in New York's 1st congressional district, the easternmost and most conservative on Long Island. In 1978
voters had elected county legislator William Carney to the congressional seat, running on New York State's Conservative Party ticket, making
him the only registered Conservative in Congress. Carney had strongly
supported the laboratory and the Isabelle project (and the Shoreham
Nuclear Power Plant), and his Conservative credentials gave him access
to officials in the new administration. Working with Carney's office
and BNL's recently named director, Nicholas Samios,[6] we organized a
visit on September 30 with David Stockman, Reagan's new director of
the budget. Accompanying Carney and me were Nobelists Yang, Samuel Ting, and Val Fitch, who was an AUI board member and helped
arrange the visit.

The meeting followed a course that, with variations, I witnessed
years later on many similar occasions. The congressman dominated
the time with introductions and advocacy remarks having little persua-
sive content. I tried to place the importance of the project in a broader
context of regional and national importance, and in the remaining
minutes Sam Ting began his presentation. Then it was over. This was
my first contact with Ting, and when he began to speak I immediately
regretted that I had not intervened earlier to give him more time. Despite my skepticism about influencing the president's budget through a

---

[6]BNL's previous director, George Vineyard, had resigned in August, and Samios
was acting director at the time of the meeting.

twenty-minute ad hoc meeting with his budget director, Ting's grave and impressive manner and his lucid and immaculately prepared presentation would have elevated the tone of the meeting and signaled the importance of the issue in a way the rest of us could not. In any case, the message from Stockman was clear. Funding for DOE's high-energy physics program would continue to erode in the fiscal year 1981 and 1982 budgets, and Isabelle was very much at risk.

Our visit with the thirty-five-year-old budget wunderkind occurred just before journalist William Greider's long article "The Education of David Stockman" appeared in the *Atlantic*.[7] Stockman's candid misgivings about the administration's vision of supply-side economics, revealed in the article, earned him "a visit to the woodshed" with President Reagan. Now a classic of journalism, the article especially interested me after meeting Stockman in person. It taught me a great deal about how Washington worked and the enormous power of an effective Office of Management and Budget (OMB) director with presidential support. The visit was my first to what is now called the Eisenhower Executive Office Building adjacent to the White House West Wing, and to the lair within of senior OMB officials. Twenty years later I would briefly occupy an office two floors above.

During the following summer, August 1982, the nation's particle physicists gathered in an historic working meeting in Snowmass, Colorado, "to assess the future of elementary particle physics, to explore the limits of our technological capabilities, and to consider the nature of future major facilities for particle physics in the U.S."[8] Organized by the Division of Particles and Fields of the American Physical Society, it was the first time representatives of all four U.S. accelerator laboratories and the university particle physics community had met to plan a collec-

---

[7]William Greider, "The Education of David Stockman," *Atlantic Monthly*, December 1981, www.theatlantic.com/magazine/archive/1981/12/the-education-of-david-stockman /5760/.

[8]Proceedings of the 1982 DPF Summer Study on Elementary Particle Physics and Future Facilities, June 28–July 16, 1982, Snowmass, Colorado, R. Donaldson, R. Gustafson, and F. Page, eds. U.S. Department of Energy and National Science Foundation. lss.fnal.gov/conf/C8206282/

THE SUPERCONDUCTING SUPER COLLIDER   37

tive future.[9] At the session on "Plans and Possibilities at the Laboratories," each lab director duly sketched his hopes, but Fermilab's Leon Lederman captured the spirit of the meeting in a visionary preamble to his own report. After describing bold European plans at CERN and Germany's DESY (Deutsches Elektronen Synchrotron), Lederman expressed his "nightmare."

> Dare we be any less imaginative? Are we settling into a comfortable, secondary role in what used to be an American preserve? . . . In the U.S., the problem is that we have, over the past two decades, been reduced to four aging laboratories. Each of these laboratories properly does accelerator R&D in order to maximize the physics that can be realized on its site. . . . In my nightmare, I noticed that none of the four labs has a large enough site [for a machine that achieves the energies necessary to see new physics]. . . . As proposals for the late '80s, all four laboratories have been pressing on projects which may not, in my opinion, provide "sufficiently bold thrusts into the unknown" and, in this sense, do not seem to me to promise to provide the excitement which draws the best and brightest. . . . I believe it is important to at least examine the possibility that the machine for the late '80s be, in fact, a very bold advance. We need to ask ourselves hard, introspective questions: are we, as a community, growing old and conservative, and is there a danger of quenching the traditional dynamism we have surely enjoyed in the past three decades?

Five months later, in January 1983, Carlo Rubbia and colleagues at CERN's Super Proton Synchrotron discovered the theoretically anticipated "weak vector bosons" $W^{\pm}$ and Z^{0}, keys to one of the deepest and most intricate parts of the Standard Model. Rubbia, who split his time between CERN and Harvard, received half the Nobel Prize in physics just a year later. The other half went to CERN's Simon van der Meer for

---

[9]The accelerator laboratories were BNL, Fermilab, Stanford Linear Accelerator Center, and CESR (Cornell Electron Storage Ring). Germany's DESY (Deutsches Elektronen Synchrotron) also participated.

a clever advance in accelerator technology that made the discovery possible. Lederman's nightmare seemed to augur these events.

In July, following this spectacular European success, and with the consensus formed at the Snowmass meeting, DOE's High Energy Physics Advisory Panel (HEPAP) recommended the "immediate initiation of a multi-TeV high-luminosity proton-proton collider project with the goal of physics experiments at this facility at the earliest possible date."[10] This proposal would become the Superconducting Super Collider (SSC). The Isabelle project at BNL, by then renamed the Colliding Beam Accelerator (CBA) and getting good reviews, would be abandoned.[11] DOE official James Kane told a *New York Times* reporter, "The panel found that all the interesting physics had passed Isabelle by. With the Europeans moving continuously ahead, and the Linear Collider under construction at Stanford, they really didn't see the purpose of building the machine." Sunk cost: 200 million 1980 U.S. dollars.

Physicists at BNL and Stony Brook were dejected by the decision, which nevertheless did not come as a complete surprise. In his Snowmass report Samios had given a realistic assessment of the situation, saying, "I noted earlier the deteriorating budget which has caused the present difficulties so that it seems prudent to us to look at ways to reduce the cost of this project," and he proceeded to describe four lower-cost projects that would exploit the new technology developed for the CBA. One of the four was a heavy ion collider—a concept that later grew into the nation's most powerful tool for high-energy *nuclear* physics (as distinct from the higher-energy *particle* physics). How Samios through vision and effective leadership resurrected the CBA carcass in a new and powerful incarnation is part of another story (Crease 2008).

---

[10]TeV is the scientific abbreviation for= Tera-electron Volt, equal to one trillion electron volts, a measure of energy.

[11]HEPAP acted on the recommendation of its subpanel on long range planning, chaired by Stanford's Stanley Wojcicki, that met June 5, 1983, in Woods Hole, Mass. Although the subpanel was "deeply divided," the HEPAP decision was unanimous.

Fermilab

Located thirty-four miles west of Chicago, the Fermi National Accelerator Laboratory (FNAL, or Fermilab) was an early offspring of a 1950s technology breakthrough pioneered at BNL. In the early 1960s two machines based on this "strong focusing" technology—BNL's Alternating Gradient Synchrotron and CERN's very similar Proton Synchrotron—exceeded all expectations and opened the way to a new generation of much larger devices. Responding to advice from a key expert panel,[12] the AEC decided to support two ambitious design projects for new strong focusing machines (1963), one at Berkeley's Lawrence Radiation Laboratory (LRL) for a 200-GeV accelerator to be built as soon as possible, and the other for an eventual machine with five times greater energy at BNL. The latter became the ill-fated Isabelle; the former evolved into Fermilab. Notice the decades-long time spans of these developments. (For details of the history of this period, see Hoddeson et al. 2008.)

Funding these ambitious projects depended on support from the Bureau of Budget (BOB, now Office of Management and Budget), which meant direct White House assent. The Fermilab accelerator alone (the "Main Ring") would end up costing $243 million in 1967 dollars, or more than $1.6 billion in 2012 dollars. In those days the nation's research and development budget was soaring toward its peak during the Apollo program, and critics had begun to complain about the use of public funds for areas of science that brought relatively little public benefit. Firm limits set by BOB early in the process narrowed options for the AEC, whose chairman at the time was Glenn Seaborg, a Nobel laureate in nuclear chemistry and an effective advocate for particle physics. Seaborg persuaded Congress and anxious physicists (who feared a low-cost machine would fail to make discoveries) to accept the BOB limits, and the project went forward. But cost was not the only troublesome issue.

---

[12]The panel was sponsored by the President's Science Advisory Committee and the General Advisory Council of the AEC and chaired by Harvard's Norman Ramsey, who collaborated with I. I. Rabi in the mid-1940s to launch BNL (see Crease 1999). This is the origin of HEPAP.

40 SCIENCE POLICY UP CLOSE

Other controversial issues involved where to locate the accelerator whose design AEC assigned to Berkeley, and how to govern it. Its large taxpayer-borne cost virtually mandated that it be widely available to a national user community, which went against a Berkeley tradition of close ownership.[13] And siting the device at or near an existing accelerator center was opposed by midwestern scientists, who felt their region had been slighted in the distribution of physics resources. The Berkeley group strongly resisted giving up control of "their" machine, and opinions within the physics community diverged widely not just on governance but even regarding the value of the entire enterprise, given its cost. These divisions threatened to undermine support from Congress, whose concern had already been expressed in AEC budget hearings conducted in 1964 by the Joint Committee on Atomic Energy. The same issues of cost, governance, location, and scientific value emerged again more than two decades later with the SSC.

Frederick Seitz, who in 1962 had become president of the National Academy of Sciences, cut through the bickering about governance by simply forming a new corporation using academy resources and inviting the presidents of twenty-five major research universities to a meeting (January 1965) to discuss plans for a future national accelerator laboratory. His idea, which had been discussed and supported by most of the players, was that the universities would form a consortium similar to BNL's AUI, but much larger. Seaborg and presidential science advisor Donald Hornig endorsed the move, and within a year the Universities Research Association (URA) had been incorporated, and trustees, chairman, and president selected. The board of trustees met for the first time in December 1965. Little more than a year later, the AEC contracted with URA to design and eventually build and operate the new laboratory. The contract to URA included design, despite the fact that the LRL design group had been hard at work on a 200-GeV ma-

---

[13]The sense of ownership was strong: the basic concept of circular accelerators was developed before and during WWII by the entrepreneurial Berkeley physicist Ernest Orlando Lawrence, whose vision launched government-supported "big science" at national laboratories.

chine for more than two years. Their design exceeded BOB cost limits, however, and their resistance to alter it weakened their political influence on the terms of the contract.

At least one congressman on the Joint Committee on Atomic Energy, Craig Hosmer (R-California), doubted that "a group of university presidents or a group of high-energy nuclear physicists are perhaps the best people in the world" to determine management arrangements for the huge new facility. Seaborg pointed to the existing AUI setup at BNL, which seemed to work well and provided for a vibrant user community. To address concerns, the bylaws of the new corporation were designed to include "principal administrative or corporate officers," as well as scientists from member institutions in its structure. Thus, URA had two layers (later changed to three in the SSC era): a Council of Presidents of the university "shareholders," and a board of trustees that does the corporate business, including selecting a full-time president, negotiating contracts with the AEC (since 1977 the DOE), and appointing the laboratory director.

Deciding where to build was trickier. The initial contention among East Coast (BNL), West Coast (LRL), and Midwest science groups was overwhelmed by the interest of numerous other parties. Following the AEC's public announcement of a June 1965 deadline for proposals, the commission was flooded with 126 responses for about 200 sites in 46 states. These numbers made it inevitable that the selection process would have a political flavor. After many months of deliberation that included a National Academy site advisory committee, the commission announced the selection of Batavia, Illinois, in December 1966. Part of the delay was caused by an NAACP objection to Illinois's failure to address discriminatory housing practices, a significant issue as President Johnson garnered support for his Great Society initiatives. His ally in this historic endeavor was Illinois senator Everett Dirksen, a Republican and author of the Fair Housing Act of 1968. Politics aside, the lengthy site selection process certainly increased support for the project, both in the public and at the White House. It deflected attention from cost and other issues and created a sense of inevitability that made it easier for the president to support the project and to integrate it into his own political plans. Within the scientific community, however, the selection process and its outcome contributed to divisions that clouded the dawn of the grand enterprise.

One month after the site announcement, the AEC awarded URA a temporary contract to launch the new laboratory. The task of finding a director fell to Norman Ramsey, who had been elected president of URA a few months earlier. It was not made easier by disgruntlement at the two older labs, particularly LRL, whose former dominance of accelerator science and technology was clearly threatened by the creation of a big new competitor. Relations with the Berkeley group worsened when URA appointed Robert R. Wilson as the new laboratory director. Wilson had been a rebellious protégé of Berkeley's Ernest Lawrence and had criticized the LRL design in 1965 for being too conservative and expensive. By 1967 he had just led the successful construction of an electron accelerator at Cornell, funded by the National Science Foundation, and had his own out-of-the-box ideas about how the nation's next large machine should be built. Seaborg, who owed his Nobel Prize to work at LRL's accelerators, supported the Berkeley concept until it became clear that budget pressures would force a reduction in scope. LRL director Edwin McMillan, who shared the 1951 Nobel with Seaborg, regarded AEC's rejection of Berkeley's site, design, and governance concepts as a betrayal. He is quoted as remembering "feeling that the AEC was determined not to give them anything." And this quote from URA president Ramsey, who visited LRL just after the site announcement: "Although I had many personal friends at Berkeley, I have never spoken to such an unfriendly audience. . . . Among other things I was told that if URA were a responsible organization it would refuse to manage the laboratory in Illinois and instead should investigate the AEC to determine how it could possibly have made such an incredibly bad decision" (Hoddeson et al. 2008).

Wilson turned out to be a brilliant choice as the new laboratory director. Over the next five years, he created a bare-bones facility on schedule and under budget. "Something that works right away is over-designed and consequently will have taken too long to build and will have cost too much," he said in 1966. His Spartan accelerator facilities

THE SUPERCONDUCTING SUPER COLLIDER   43

were more than compensated, however, by his insistence on an imaginative aesthetic dimension for the lab site, informed by his own architectural instincts and his considerable artistic talent. Fermilab became a model of environmentally sensitive campus planning and architectural innovations unusual even by today's standards for buildings at national laboratories. The site is dominated by a cathedral-scale structure chosen and partially designed by Wilson. Wilson's own sculptures dot the landscape, and a small herd of bison grazes in the 800-acre prairie preserved in the center of a berm over the Main Ring. The ring itself lies in a tunnel whose dimensions, at Wilson's insistence, allowed for the later installation of the superconducting ring that became the Tevatron.[14] The date of Wilson's resignation in 1978 is significant in the larger scale of national events, as I explain below.

URA in the Superconducting Super Collider Era

Following its action on Isabelle, the DOE immediately launched plans for the vastly larger machine that would become the SSC. URA's large regionally balanced membership and flexible corporate structure made it a logical conduit for funds and oversight for the new activity, and it was through URA that I became involved with the project.

By the mid-1980s few URA member presidents elected to attend the annual Council of Presidents meetings. They delegated other officials—research vice presidents, deans, or senior faculty—who were closer to the physics issues. Presidents who did appear, usually the day before the meeting, were treated to a small dinner with corporate officers at the Cosmos Club. The URA leadership welcomed my participation, and I learned much through these informal sessions. As a young president of a relatively new university, I found URA an efficient entry to Washington

---

[14]Hoddeson et al. (2008) sketch Wilson's biography, but his remarkable personality and the Leonardo-like breadth of his accomplishments deserve an extended volume for a broader audience. His own personal account, "Starting Fermilab: Some Personal Viewpoints of a Laboratory Director (1967–1978)," can be viewed online as a Fermilab "Golden Book" in the lab's History and Archive Project website at http://history.fnal .gov/GoldenBooks/gb_wilson2.html.

circles that were important to Stony Brook's long-term research interests. I became chairman of the Council of Presidents in 1987, and my active engagement began just as the corporation was reorganizing itself to avoid conflicts of interest in the SSC site selection process. The new board of trustees elected me chairman early in 1988, and I served in that capacity through the entire construction and termination phase of the SSC, until I became director of BNL in 1997. It was on the whole a painful experience, relieved only by the actual remarkable progress in the main business of the SSC lab (SSCL) despite all obstacles, and by the continuing successes at Fermilab, the SSC's older sibling under URA, capably directed by Leon Lederman (Nobel Prize 1988) and, after July 1, 1989, by John Peoples, Jr.

For five years URA struggled to launch the new SSCL in the face of increasing difficulties, until Congress canceled the project in 1993. By then it had dawned on me that cancellation was a likely outcome, so I was emotionally prepared for what was a cataclysmic event for the physics community and thousands of SSCL employees.[15] Other things did take me by surprise, however, that opened my eyes to troubling aspects of "big science" management by federal agencies. These became important later, as I explain in the following paragraphs. The complex history of this monumental project divides into two sharply contrasting epochs, of which I was fully engaged only in the second. The brief account below of the first draws on my own later experience, with some assistance from supplemental sources.[16] And yet there is a larger, simpler view that benefits from hindsight: the SSC advocates, myself included, misjudged the value society would place on the scientific product and pressed forward a machine whose cost simply outweighed the benefits as perceived by the public. Elected officials tolerated the expense in the early years and fostered hopes, as politicians do, that the project could eventually be funded. When the time came for big increases

---

[15]Herman Wouk's novel *A Hole in Texas* (2004) paints a picture of aspects of the impact of the SSC's cancellation on employees.

[16]These other sources are the detailed chronicle of Hoddeson and Kolb (2000) and an important eyewitness account by Stanford's Stan Wojcicki (2008, 2009). I will not repeat the details here, but commend readers especially to Wojcicki's excellent papers.

THE SUPERCONDUCTING SUPER COLLIDER   45

to maintain construction schedules, the political support died away in a
messy fashion whose details are less important than they seemed at the
time. A few details are necessary for a coherent account, but my interest
here is on larger policy implications of the entire experience.

The SSC Design Phase

The postwar accelerator boom produced a cadre of accelerator builders,
including physicists, engineers, technicians, and other skilled workers.
They worked in teams to produce exquisite instruments—huge detectors,
as well as accelerators—on scales of size and complexity unprecedented
in the history of science. Like other scientific communities, they published their ideas in professional journals and met in international conferences to discuss and plan for future projects. Ideas for a super accelerator had been incubating in this community for years when the 1982
Snowmass meeting ignited a spirit of possibility that grew rapidly into a
conviction that a super machine could be achieved. Events moved quickly
after the DOE accepted the fateful HEPAP recommendations in late 1983.

That December DOE reprogrammed modest resources to support
a "Reference Design Study" needed for a preliminary cost estimate. A
team of about thirty emerged quickly under the leadership of Cornell's
Maury Tigner, a Wilson protégé, and withdrew to Berkeley's LRL to attempt to produce a rapid but credible report. Looking back on this effort
through the telescope of my own later experience, I thought to myself,
"And then a miracle occurred." Within four months the team produced
an impressive 440-page document that became the "bible" for all subsequent SSC design work. More than a hundred additional engineers
and physicists contributed in visits, workshops, and essentially volun-
teer labor. Even accounting for the fact that the cost was dominated by
a few elements, notably the superconducting magnets, the number and
technical difficulty of details that had to be considered to arrive at a total cost are mind-boggling. Three different magnet designs were considered, all leading to approximately the same overall cost.

The Reference Design Study cost estimate was about $3 billion in 1983
dollars, just to build the facility. In this and later reports the authors

emphasized what features were and were not included in the estimate. Costs were so important to success, however, that even more emphasis was warranted. In retrospect, it was a strategic error not to highlight the total costs—preconstruction R&D, site acquisition, energy, contingency, detectors, and so forth—in the very first estimate. When these extra costs appeared years later in various government documents, they played on the political scene as cost overruns and contributed to a general image of mismanagement.

With the Reference Design Study in hand, DOE launched the next phase by establishing a funded Central Design Group (CDG) under contract with URA, expanding the effort at the LRL location to about 250 people, who delivered a full report in 1986. Tigner was the natural choice once again to lead the expanded team. The dry words "Conceptual Design Report" do not convey the importance of this phase of the enormous enterprise. The SSC circumference was about the size of Washington, DC's beltway, and the accelerator feeding its main ring would itself be larger than any previous device. Three additional accelerators would be required to bring the proton beams up to the energy and quality needed at the final stage—all this to examine nature's smallest parts! The protons would circulate in bunches about a millimeter in diameter, focusing down to microns at the collision points. How to steer hair-thin beams over miles-long distances without dissipating or crashing into the walls of the guiding tubes is a major technical challenge. Producing a detailed design (not yet a full design before the site was known) for a project of this scale in little more than two years was a phenomenal feat. The group commissioned substantial original research to develop concepts and components, especially the superconducting magnets that would have to achieve unprecedented levels of performance. Much of the work was farmed out to the other DOE labs, universities, and industry, but it had to be conceived, coordinated, and integrated into a coherent and exquisitely articulated functional whole. Tigner was the hero of this effort: "He set the tone for the organization, provided inspired leadership, and set an example by working long days, frequently seven days a week. CDG goals were achieved with a minimum amount of bureaucracy and in a stimulating intellectual atmo-

THE SUPERCONDUCTING SUPER COLLIDER   47

sphere" (Wojcicki 2008, 272). The other leaders indispensable to this effort were Chris Quigg, David Jackson, and Stan Wojcicki. I first met Tigner at a URA board meeting and was impressed with his direct and unassuming manner and complete command of the details of the awesomely complex SSC project. Many years later I remarked in an oral history interview that "Maury is—was then, is now—regarded as one of the key accelerator physicists of our time. He had a good reputation and was a logical person to ask [to direct the CDG]. There aren't too many of them."[17]

Public awareness of big projects normally begins when bulldozers arrive and steel goes up. But the work on the SSC depended utterly on the talent of highly skilled professionals who labored less visibly to translate a vision into the specifications and functions of a staggering number of components. The translation process is particularly difficult for complex scientific instruments because the whole point of building one is to exceed the power of all its predecessors. That is why key components are designed initially by scientists. They have access to emerging knowledge of materials and techniques not yet incorporated in engineering practice. Each new accelerator is a new invention and a test bed for previously untried materials and technologies. The designers try to see into the future to estimate the state of the art when it is needed in the construction cycle, not as it exists "off-the-shelf." They expect the instrument to be used and maintained by scientists and engineers who will make innovative upgrades to keep up with the expanding frontier of discovery. When the project is very large, this expectation leads to a deep and difficult dilemma, already evident in the skepticism expressed by Congress in 1964 about whether scientists and academics "are the best people in the world" to manage such projects.

Scientists point to the wartime Manhattan Project as the paradigm for their role in producing "big science." That effort was unique in the quality of interaction between the scientists and their military customer.

---

[17]James H. Meredith Oral History interview, June 6, 2009, conducted by Lillian Hoddeson and Michael Riordan.

Its successful outcome—if "success" can describe a weapon of mass destruction—depended on wartime conditions and the leadership of two extraordinary men: General Leslie R. Groves, Jr., and J. Robert Oppenheimer. After the war, however, relations stiffened rapidly into more normal bureaucratic patterns, and early AEC-sponsored large-scale scientific projects encountered what might be called "strategic management difficulties." Government agencies attempt to ensure what Congress perceives as effective use of public funds. Industry possesses the organization and experience needed for the management of large construction projects. Scientists have the knowledge and skills indispensable for deep troubleshooting through all phases of design, construction, and operation. Success requires a level of mutual understanding among all three cultures that is very hard to come by. Early attempts to speed the installation of new instruments through "turnkey" procurements from industrial suppliers often failed at BNL because none of the parties understood how to deal with each other on these unique projects (Crease 1999).

A shared sense of national urgency at war's end brought the disparate parties together in working arrangements that gradually built up competencies within each sector. Europe and the Soviet Union also had communities of accelerator scientists, but the United States developed governance and procurement models that preserved the prewar lead in experimental particle physics established by Ernest Lawrence and the Berkeley group. The resulting series of accelerators at Berkeley, BNL, Stanford, Cornell, and Fermilab produced most of the discoveries that led to the Standard Model. Other nations copied the U.S. model of government-owned laboratories operated by quasi-independent contractors with multi-institutional governing bodies. With time, however, other "exogenous factors" intervened in the highly successful AEC/university/industry relationship that eroded its effectiveness. Recessions, hot and cold wars, geopolitical crises like the Arab oil embargo and the Iran hostage crisis, and the unpredictability of the U.S. political scene all had their effects on the course of postwar science.

Two factors of very different kinds, however, emerged during the 1970s and grew during the 1980s to have a persistent and dominating impact on the DOE national laboratories: the changing government interface with accelerator physics, and accountability, or what I'll call the Principle of Assurance. These two factors converged just as the SSC entered the construction phase, and their combination produced the messiness, not to say the ugliness that I mentioned above, in the final years of the project.

The Changing Government Interface with Accelerator Physics

The episodic transformation of postwar federal administrative structure and accompanying management reforms were an important factor in the trajectories of DOE national laboratories during the 1980s. A growing commercial nuclear power industry made it desirable to separate the regulatory from the operational functions of the AEC, which had inherited the entire nuclear portfolio from the Manhattan Project. This occurred during a transition period in the 1970s when Congress created the Nuclear Regulatory Commission and combined the AEC with the Energy Research and Development Agency, an outgrowth of earlier executive branch energy offices. The result was a very much larger and more bureaucratic cabinet-level DOE that combined the nuclear weapons program, the operations of national laboratories, and a wide array of energy policy, regulatory, and applied research programs. Seaborg resigned near the end of the AEC era in 1971, after a decade of distinguished service. His successor, James Schlesinger, had a powerful impact on the evolving agency, introducing key organizational and management changes.

Strong-minded and hawkish, Schlesinger became secretary of defense under Nixon, was dismissed from that post by President Ford in late 1975, and was then appointed by President Carter to be the first secretary of the new DOE (1977). Two years later Carter replaced him with businessman Charles Duncan. Whatever Carter's political reasons were for bringing in the prickly Republican, it is clear that Schlesinger's aggressive and demanding style would be an asset in implementing the "zero-based budgeting" management approach Carter advocated. Consolidating the host of programs in the new department was a major task that fit Schlesinger's experience and insistence on results. During this historic decade of national change, the federal interface with big physics fragmented and grew thick with bureaucracy. The technically savvy and supportive partnership of the Seaborg era would never return, and subsequent government reform movements, culminating in the Government Performance and Results Act of 1993, created serious new challenges for large-scale scientific projects.

As Fermilab's Main Ring began to operate early in this transition decade (March 1972), Wilson launched his campaign to secure funding for an additional accelerator ring in the same tunnel using superconducting magnets. This would save electrical power and double the working energy of the accelerated protons, hence the name "energy doubler" for the project. BNL was also designing superconducting magnets for Isabelle at this time. Wilson was frustrated with the new bureaucracy and the dispersion of congressional authority for different pieces of the emerging energy agency apparatus. Funding difficulties in the recession-ridden 1970s only made things worse, and Wilson's dissatisfaction with DOE decisions to keep the Isabelle project going in competition with the doubler eventually led him to resign as lab director in 1978. The doubler was completed under Lederman, and its big detectors, named CDF and D0, were assembled in parallel with the SSC design phase.

Wilson's campaign for the doubler coincided with the introduction of two major government-wide management reforms: Nixon's management by objectives initiative (1973) and Carter's zero-based budgeting initiative (1977). These were preceded in 1965 by Johnson's program planning and budgeting systems reform. "Oversold and poorly designed, those management reforms were unrealistically scaled and difficult to sustain and advance. Largely as a result, they died relatively quickly, only to be reinvented in a seemingly inexorable cycle" (Forsythe 2000, 11). By the end of the decade, however, it was clear that Japan's adoption of similar management practices in the private sector created a competitive advantage that led to serious losses of U.S. manufacturing market share, especially in the automotive industry. U.S. manufacturers belatedly rediscovered and applied these practices during the 1980s. Their evident success encouraged public administrators to adapt them to government operations as well, "inexorably" resurrecting features of the 1970s executive branch reforms. By the end of the 1980s the national laboratories were experiencing the reform movement through increasing demands for accountability in their contracts with DOE. Tensions rose as the contractors—most of them universities or coalitions of universities, like URA—strongly resisted provisions that would increase liability for laboratory operations and sanctions for regulatory infractions.

These issues emerged at the February 1987 URA Council of Presidents meeting where I presided as council chair. President Reagan had recommended the SSC to Congress a month earlier, but a full year after the Conceptual Design Report had been completed, and URA was eager to secure an arrangement with DOE to launch the project. Ed Knapp, URA's president throughout the Reagan era, requested an endorsement from the members for an unsolicited URA proposal to build and operate the SSCL. I asked Knapp to address the issue of university management of the SSCL, which was central to URA's unsolicited proposal. In the ensuing discussion, MIT provost John Deutch, who among many other roles had served as Schlesinger's Director of Energy Research in 1977–1979, remarked that "there are several obstacles in the federal government to choosing university management for the SSC: (1) the government needs to demonstrate competitiveness in contracting; (2) DOE operations offices think of themselves as the actual contractors, with the university groups only in a supporting role; (3) there is a perceived need for a strong systems integration component in the management of SSC construction, which a university group may not be able to handle."[18] And indeed, the contract DOE awarded to URA two years later, following a formal and open competitive procurement, included management requirements unprecedented in the history of accelerator construction. Neither DOE nor the accelerator community was prepared to implement these requirements. They were transmitted informally but insistently to URA throughout the design phase and formally embedded in DOE's eventual request for proposals (RFP) for the SSCL—and they led URA to seek industrial partners who could enhance the management credibility DOE demanded.

---

[18]Minutes of the Annual Meeting of the Universities Research Association Council of Presidents," February 27, 1987, National Academy of Science, Washington, DC, p. 6. In the John H. Marburger III Collection, Box 15, "URA 1988" folder, Stony Brook University Special Collections and University Archives (note added by editor).

As URA chairman I met frequently with the working group headed by BNL's Martin Blume that prepared the proposal during 1988, and I puzzled with the others over the strong messages from DOE. URA's unsolicited proposals (March 1987, February 1988) met with silence, and DOE threw open the competition to industrial bidders, signaling a determination to bring industrial "best practices" to bear on SSC construction. The RFP went out on August 22, with responses due November 4, just before the national election that swept Reagan's vice president, George H. W. Bush, into office. Knapp recruited N. Douglas Pewitt, who had been deputy to Reagan's presidential science advisor George Keyworth in the Office of Science and Technology Policy and later acting director of DOE's Office of Energy Research, to advise on the formal preparation of the necessarily long and complex proposal. Some commentators have attributed URA's decision to engage industrial partners to Pewitt's influence, and that is partially correct. Pewitt seems to have been the only one in the group who fully appreciated the "inexorable" momentum of contract reform within DOE. He also understood how DOE would evaluate the proposal through a line-by-line comparison with the RFP requirements. Wolfgang Panofsky, who had been director of SLAC until 1984 and had the deepest laboratory management experience in the group, viewed the RFP requirements as highly inappropriate and certain to lead to trouble as the project got under way. All of us concluded, however, that a failure to name industrial partners would either render our proposal noncompetitive or lead to a drawn-out contract negotiation in which we would be forced to accommodate industrial participation one way or another. Panofsky and Pewitt wrote fascinating back-to-back letters for the March 1994 issue of *Physics Today*—after the project was dead—that reveal sharply contrasting views of the significance of DOE's demands.[19] Panofsky wrote that the reason URA was forced into this arrangement was "to broaden support in Congress." Pewitt wrote that "Congress only serves as a jury, and physicists

---

[19]Wolfgang K. H. Panofsky, Doug Pewitt, David R. Nygren, Pierre Ramond, Robert J. Reiland, Christopher Carone, and Rustum Roy, "The SSC's End; What Happened? and What Now?," *Physics Today* 47 (March 1994), 13, 15, 88–89.

THE SUPERCONDUCTING SUPER COLLIDER   53

are not exempt from the bureaucratic procedures that are now part of all large publicly funded projects." On this point Pewitt was right and Panofsky was wrong. But Panofsky's prediction that the requirements would lead to trouble were accurate. His *Physics Today* letter recounts some of the consequences that I describe below.

URA's proposal process has been portrayed somewhat critically else-where as departing radically from historical transactions with govern-ment that launched previous big accelerators (see below). This is true but misleading to the extent that URA—or the accelerator community—might be seen as having any choice in the matter. Nor were URA's choices of industrial partners motivated by a desire to capture their "po-litical clout" for the project (but this is certainly true of choices for in-dustrial members of URA's various boards). Fragments of the old AEC partnership model that lingered in DOE's Office of Energy Research were being transformed "inexorably" by the broad public accountability movement. That movement was about to be reinforced by the second exogenous factor I mentioned above, but more needs to be said at this point about public accountability.

The Principle of Assurance

The RFP requirements were clearly intended to increase DOE control over contractor performance through, among other things, the introduc-tion of a kind of formal management system on a scale new to DOE ac-celerator construction projects. From the contractor's perspective the new system had superfluous features that impede efficient project man-agement. From DOE's perspective these features significantly increased the transparency of operations to DOE contracting officers and external stakeholders, including Congress. In fact modern performance-based management systems for publicly funded projects have two objectives, clearly defined in Dall Forsythe's brief but useful 2000 report on the Government Performance and Results Act: "At the heart of the perfor-mance management movement in Washington and around the world is a simple set of insights—managers cannot act consistently to improve services without data to track performance, and policymakers and citizens cannot make judgments about the relative value of government activities without measures of program performance and impact" (5). The SSCL contract required, among many other things, the implementation of a "cost and schedule tracking system" designed to serve both ends.

Roughly speaking, the idea is to build a simulation of the entire project, including costs and milestones, so that actual expenditures and progress can be compared with the model "on the same page," somewhat like double-entry bookkeeping. Whether this adds value to the project depends entirely on how the system is used, which entails two challenges. The first challenge is building the model—a difficult but low-risk exercise in a private company where deviations between the model and experience are signals to management to take some action (the *Act* in W. E. Deming's famous Plan–Do–Check–Act cycle). Management could change either the model (the work plan) or the manner of execution (e.g., reassigning personnel) to keep within overall budget and time constraints. I think of this as managing to budget and schedule, which is more or less how Wilson operated when he built the Main Ring at Fermilab. In the public sector—where DOE contractors operate despite their nongovernmental status—building the model becomes a much more difficult, high-risk exercise because the work plan is more than a management tool. It amounts to a commitment to the public— that is, to program managers, inspectors, auditors, regulators, reporters, and congressional staffers—that this is the way the work will be done. I call this the Principle of Assurance. Deviations become highly visible, and changes require approval by the public overseer.

The second challenge is to implement a fast and low-risk change control system—low risk because otherwise fear of seeking changes to the plan inhibits the feedback needed to drive management improvements. In the public sector the concepts of "control" and "change" are antagonistic, and "change control" can become an oxymoron, leading to what I call "managing to the model." This mode always incurs larger overhead expenses and contingency pools to cover the inevitable cost overruns and schedule slippage because of the lengthened approval process. Far from saving money, even effective implementation of the

THE SUPERCONDUCTING SUPER COLLIDER 55

Principal of Assurance in the public sector can add significantly to project cost and risk. With time DOE and its contractors (or at least most of them) learned to live with the accountability aspect of public sector performance-based management. Today's powerful information technology and growing cadre of professional project managers ease the pain, but the unpredictable consequences of broad transparency— "working in a fishbowl"—continue to be a major source of risk.

As the SSC project proceeded from design to construction, the demands and side effects of this highly public accountability machinery were encountered for the first time by the accelerator builders, and implementation of a rigorous cost and schedule tracking system fell far behind the expectations of federal overseers—some of them, at least.

At the time not all overseers agreed on the significance of the new approach to contractor accountability. Here are excerpts from the prepared statements of two witnesses at an April 1992 hearing of the Subcommittee on Investigations and Oversight of the U.S. House Science, Space and Technology Committee.

Statement of Victor Rezendes,
Director, Energy Issues, Resources,
Community and Economic Development Division,
General Accounting Office (GAO)

Although DOE maintains that the SSC project is on schedule and within budget, it does not have in place an integrated system for monitoring cost and schedule performance that would allow it to objectively determine its progress. . . . [A]fter more than 3 years as the operating contractor for the SSC Laboratory, URA has yet to implement a functioning Cost/Schedule Control System. As a result, DOE lacks objective information to assess on a timely basis whether the project has encountered problems affecting its cost and schedule. . . . In the absence of a fully integrated cost and schedule system, the potential impact of all cost and schedule changes that have been made or are being considered is not known.

Statement of Robert M. Simon,
Principal Deputy Director,
Office of Energy Research, DOE

In [January 1991] DOE published the cost and schedule baseline for the project. I can report to you today that the project continues to be executed within that baseline. . . . All of the major contract awards that have been made are at or below the baseline estimates. This includes the contracts that have been let to date for major cost components of the Super Collider, including superconducting magnets, cryogenics plants, and tunnels. In this way we have validated the original estimate in the only way that is useful—in the marketplace. In a similar fashion our schedule performance is consistent with the baseline schedule for commissioning the collider at the end of [fiscal year] 1999. The project cost and schedule has been maintained with no compromise of the technical scope of the project.

Rezendes was wrong with respect to DOE's "objective knowledge to assess" but right in the perception of a lack of public transparency associated with the absence of the required cost and schedule control system. The phrase "potential impact . . . is not known," however, is a matter of interpretation, and the gap it opens between the GAO interpretation and that of DOE or URA here is significant. Two weeks after the hearing, *Chronicle of Higher Education* reporter Kim McDonald interviewed SSCL's general manager, Ed Siskin, regarding the hearing issues. Among Siskin's lucid and detailed responses recorded in URA's transcript of the interview is the remark:

I believe we have an integrated cost program. I believe it is not as automated as people would like and it won't be as automated as people would like before the end of June. But the information is still there, the comparison of the actuals to the baseline amount still is made on a monthly basis. A variation analysis is still done on a monthly basis. When I've asked GAO or DOE is there anything else that I ought to be doing in their view, the answer is no. I don't know what else to ask.

THE SUPERCONDUCTING SUPER COLLIDER

The only thing that they were concerned about was that the same piece of paper does not show both the monthly expenditures and the baseline amounts. It takes two pieces of paper. I acknowledge right now that that's the case.

Many incidents hostile to the SSC flowed from these hearings, which hinged on issues that would have been minor in the private sector but had disproportionate impact in the public domain. Another exchange in the *Chronicle* interview transcript illustrates the risk to the contractors of operating in this completely transparent way:

McDonald: GAO contends that moving the detector halls could generate as you heard in the hearing a 13 month delay adding dollars in cost.

Siskin: Let me explain . . . GAO is essentially in residence as is IG [DOE's Inspector General] and about everyone else that might choose to want to be here. When you make any change in a program whatsoever, the first thing you do is put in that change and see what your cost control and schedule control system kick out. And that way you know what things you have to address and what things you have to resolve. One of the first indications when we put in the change in direction [for the detector halls] . . . kicked out a draft possible effect of 13 months. It stayed under study for a few days before ways were recognized to reduce it. The present completion date in the schedule is . . . six months ahead of completion of the collider. We are caught betwixt and between. We're trying to be completely open, people are seeing access to our internal working documents the first time they are kicked out. But certainly it requires the ultimate in unreasonableness to say that the first time you do a study to identify where potential problems are, that represents the reality. That's nonsense. That's pure gibberish.

None of Siskin's extensive explanations to McDonald appeared in print, and the resulting article was hostile.[20] The combination of rigorous enforcement of detailed management processes like the cost and

---

[20]Kim A. McDonald, "Investigators Criticize Management of SSC," *Chronicle of Higher Education*, May 6, 1992.

58   SCIENCE POLICY UP CLOSE

schedule control system and a workplace flooded with external observers with multiple agendas creates heavy risks for the contractor. Mitigating that risk requires sophisticated enterprise software systems, a large professional staff to maintain them, and extensive (usually unpopular and time-consuming) training for all employees who are likely to come into contact with the armies of external inspectors, auditors, and others who are attracted by the political salience of expensive public projects. Luck, too, plays a role, and the SSCL had little of it in its phase of the decade-long saga of the SSC.

SSCL Proposal and Handoff

URA members pay no annual dues but agree to occasional assessments up to a cumulative maximum of $100,000. This was the source of funds needed ($600,000, or about $1M in 2010 dollars) to cover the cost of a formal management and operating proposal responsive to DOE's enhanced version of the federal acquisition regulations (introduced in 1984), a first for URA or any other national laboratory contractor. Much of the proposal content came from the Central Design Group (CDG), but the frenzied writing and testing of the document in the seventy-four days allotted by DOE was done by the smaller group of corporate officers, consultants, and members of the CDG/SSC board of overseers (BOO). DOE was actively encouraging other bidders ("the government needs to demonstrate competitiveness in contracting"), and it was necessary to work with a secrecy alien to the wide-open traditions of the accelerator physics community. Lillian Hoddeson and Adrienne Kolb's 2000 article gives an important and accurate account of the process based on interviews with its key participants, but their report is filled with references to "beltway bandits," "hired guns," and editorial choices of remarks by the interviewees that suggest URA descended to distasteful practices to win the contract. I viewed the process as business-like in a way unfamiliar to a science community whose own approach to business was extraordinary. Hoddeson and Kolb's account unfairly demonizes the consultants without whose contribution the subsequent negotiations with DOE would have added painful and frustrating

THE SUPERCONDUCTING SUPER COLLIDER   59

months to get to the same point. These biases may be a consequence of trying to place the episode in the "frontier" metaphor favored by the authors. The result is a narrower focus on URA decision making that overlooks a much deeper context of changes within the executive branch of government and particularly within DOE. Hoddeson and Kolb are entirely accurate, however, in their observation that "the exclusion [of the CDG from the process] harbored lasting resentment." URA's corporate management of relations with CDG, repeatedly touted in the proposal as its most valuable asset, was seriously flawed.

With a Council of Presidents, a board of trustees, and two BOOs, each having a chair and an executive committee, URA had an impressive machinery to govern Fermilab and the CDG. Its then sixty-six members included most U.S. universities offering physics doctorates, and its corporate board of trustees included industry CEOs whose experience complemented that of the academics. Distinguished members notwithstanding, the tiny URA corporate office in Washington had to rely on its laboratories for deeper management support. During the SSC design phase this situation led to asymmetrical treatment of Fermilab and the CDG, to the disadvantage of the latter. CDG was a squatter at a non-URA lab (Berkeley), remote from Washington and Illinois, with limited control over inadequate funds but focused intensely on its single complex mission. Fermilab was a going concern with fully staffed administrative departments and a constant strategic awareness of its immediate scientific future and of the Washington scene. Fermilab's BOO consisted of the former URA board of trustees. SSC's BOO, chaired by Tigner's Cornell mentor, Boyce McDaniel, was a recent invention that was preparing for a new lab, a venture that diluted attention to the CDG. Not that CDG seemed to need trustee level attention, given Tigner's effective leadership. In the end, however, URA botched its strategic management of CDG by taking it for granted.

This happened early on my watch, before I realized there was a problem, and I failed to catch it because I was detached from the social network in the accelerator community. Like other trustees, I focused on national issues and the evolving SSC political context, leaving the highly qualified BOOs to provide expert oversight. When the SSC BOO proceeded to choose names for the new SSCL director (needed for the proposal), it failed to consult adequately with the CDG community. It selected Roy Schwitters rather than one of the existing URA directors (Lederman or Tigner) in a rushed process that lacked transparency and bred suspicions about conflict of interest. Wojcicki's bitter but fair account (2009) is well supported by documents and interviews and my own brief experience at the very end of the process. As the search wound down toward the end of August 1988 (Schwitters was chosen at a meeting held at O'Hare Airport on August 28, 1988), McDaniel called to brief me in a worried tone. The search schedule had slipped; the inputs on candidates were bathed in rumors derogatory to all, many of which were attributed to political or DOE figures; Panofsky, the strongest personality on the BOO, was known to be Schwitters's strong advocate; and McDaniel, a man of the highest integrity, felt a keen personal responsibility for letting things get out of hand. His mood depressed me because I knew how fault is dispersed within organizations under stress, and how little control he could have under extremely adverse conditions. Panofsky informed Tigner of the decision that weekend, but the choice of director was confidential to the proposal, which was locked up for two more months. When finally the URA leadership—McDaniel, Knapp, and I—traveled to Berkeley to brief the CDG on the proposal, emotions were running at a high pitch.

Years later when I read Hoddeson et al.'s 2008 book on the history of Fermilab, I was stunned by the symmetry between that meeting, in mid-September 1988, and the hostility Norman Ramsey had encountered on a similar trip to the same destination, at the end of 1966, to explain the choice of site, director, and a loss of design control for what was to become Fermilab. Twenty-two years later the three of us "from URA" met whatever CDG staff could fit in a conference room, with our backs against the wall, struggling to articulate and justify the proposal process. The room became smaller as we each said our piece. Twenty years after that 1988 meeting, in my interview with Hoddeson and Michael Riordan, I could recall little of what was said: "I wasn't prepared for the intensity of the emotion that I saw there, but I thought that the emotion itself was a sign that there was something wrong. . . . For ten-

THE SUPERCONDUCTING SUPER COLLIDER   61

sion like that to build up, there would have had to be something rather wrong, and clearly the subsequent interaction between the CDG and Maury on one hand, and Roy and the [SSC] lab on the other, revealed deep, deep fissures in the relationship."[21] The proposal went in on November 4, listing Tigner as deputy director, but the deep fissures persisted as Schwitters sought to negotiate a deputy role that both could accept. "We discussed project manager," said Schwitters in a later interview, "we discussed deputy, we discussed all kinds of things. I thought Maury was very difficult to talk to." Four months later Tigner resigned from the project. The entire episode, the scale of the enterprise and its prideful characters, had a foreboding, Homeric quality—the stubborn Agamemnon seizing the prized Briseis from Achilles to launch the disaster of the *Iliad*.

DOE Begins to Manage the SSCL

"DOE operations offices think of themselves as the actual contractors, with the university groups only in a supporting role." That remark, by John Deutch at the 1987 Council of Presidents meeting, was far deeper than I realized at the time. DOE, the world's largest sponsor of physical science research, administered its programs until 1998 under an array of divisions reporting to a director of energy research (now Office of Science), which in turn reported to a DOE undersecretary. The CDG activity fell under the Division of High Energy and Nuclear Physics, experienced in accelerator construction. To prepare for its oversight of the SSCL management and operating contract, not yet awarded to URA, Energy Undersecretary Joseph Salgado created a new organization within DOE in parallel with the existing Division of High Energy and Nuclear Physics, but still within the Energy Research directorate. This action separated DOE's SSCL management from the office that had worked with the accelerator community over decades of accelerator building experience. New units were created within the new Office of the Superconducting Super Collider (OSSC) corresponding to each of

---

[21]Interview, June 6, 2009.

the major operating divisions that URA had proposed for the new laboratory, creating a duplicate "shadow" management structure that never worked as intended.

As URA prepared its proposals, solicited and unsolicited, the question of roles dominated discussions of governance and key personnel to lead the new laboratory. URA made a systematic attempt to avoid conflict of interest in the forthcoming site selection process by forming separate BOOs for Fermilab and the CDG. The latter would presumably merge into the SSCL, giving both BOOs substantial independence under the overarching board of trustees and corporate office. After the contract was awarded, Fermilab chafed briefly under an early effort by Knapp to recruit advocates for SSCL funding, apparently in competition with Fermilab's next big project—the Main Injector. The trustees subsequently declined to endorse a request from the Fermilab BOO supporting the Main Injector (a request that was made somewhat undiplomatically the day after DOE announced Texas as the preferred site). I defined internal URA roles in a white paper that compared URA with a multicampus university system, where the lab directors were like campus presidents who operated more or less independently under the corporate guidance of their respective BOOs. The trustees and corporate office would not take sides but would support a strong U.S. high-energy physics program. Knapp's successor, Johnny Toll, whom I recruited when Knapp returned to Los Alamos in 1989, was a strong advocate for both labs, and I credit him with working hard to maintain balance between the two, and particularly with keeping the Main Injector on the radar screen in his vigorous advocacy for the overall high-energy physics program.[22]

---

[22]My career has been amazingly entangled with John S. Toll's, who had been my predecessor as president of Stony Brook University (1965–1978) and, even more remarkably, chaired the University of Maryland physics department and judged a statewide science fair there during the 1950s in which I won second prize for a project on the effect of a magnetic field on the hydrogen spectrum. His Princeton advisor, Nobelist John Wheeler, had this to say of him, among other favorable comments: "While working in one of my favorite areas of mathematical physics, the relation between the scattering of particles and their absorption, [Toll] obtained elegant new results on what

THE SUPERCONDUCTING SUPER COLLIDER 63

The URA boards actively monitored the creation of these internal arrangements and maintained close contact with the other national laboratories to strengthen our negotiating position with DOE in the early months of 1989. We strongly resisted the introduction of contract language that would undermine what we saw as essential scientific authority at the other labs, including Fermilab. We finally agreed to a contract with language that we felt gave the necessary independence, but only after receiving many reassurances from the contracting officer that certain ambiguities would be worked through in consultation with URA management, and began to carry out the steps toward building the SSCL that had been carefully worked out (for a generic site) by the CDG.

DOE's decision to separate its internal OSSC from its own legacy of knowledge about building large accelerators went too far in valuing theoretical "general management" ideas over concrete operating experience. Scientists may not be the "best people in the world" to manage large projects, but they are at home with complexity and tend to work in teams whose collective capability far exceeds the norm for more conventional projects. Earlier I mentioned that administrative barriers have

---

is called the 'scattering of light by light'—that is, the collision of one photon with another. Many later researchers built on this work. Toll was a thoughtful person interested in social issues and devoted to service" (Wheeler and Ford, 1998, 181). I, too, was interested in photon-photon scattering and studied Toll's papers during my research career at the University of Southern California. Toll is credited with launching Stony Brook's research prowess during the 1970s in a highly effective partnership with New York's governor Nelson Rockefeller. He returned to the University of Maryland as president and chancellor, where he served effectively until 1989. I had no trouble convincing the URA boards to invite him to be president after a brief search when Ed Knapp resigned later that year to become president of the Santa Fe Institute. After the SSCL termination, Toll continued his remarkable service as a strong president of Washington College in Maryland, retiring in 2004 to return to physics research at the University of Maryland. I encountered Wheeler as an undergraduate at Princeton seeking career advice shortly after Toll completed his Ph.D. there. My love for the fundamentals of quantum theory and my interest in public service were strengthened by those encounters, as I am sure Toll's were. The community of physics is riddled with such personal interconnections that are a huge source of internal cohesion for the field, and an asset for federal policy in fundamental physics. It is noteworthy that nearly all presidential science advisors have been physicists or physical scientists.

a disproportionate impact on large, highly technical projects with few precedents, like the SSC. In any case, the creation of the OSSC turned out to have many fateful consequences. These are well described by Stan Wojcicki's paragraph on the SSCL-DOE relationship during this period (Wojcicki 2009), which is accurate according to my personal notes. I replicate a very condensed version of his account below because 1989 was a crucial year not only for the SSC but for the entire future of the DOE's management of its national laboratories. It was the year in which the "second exogenous factor" emerged dramatically to alter the management environment.

Initial arrangements for DOE's SSCL management were made after Alvin Trivelpiece resigned as director of energy research in April 1987. Although President Reagan shortly thereafter announced his intention to nominate Robert O. Hunter as Trivelpiece's replacement, the appointment did not go through until more than a year later, the position being filled on an interim basis by his deputy, James Decker. When Hunter finally arrived toward the end of 1988, he confirmed and strengthened the OSSC arrangements, and DOE launched a sequence of SSC actions that began with the announcement of Texas as the preferred site on November 10 and ended with the designation of URA and its industrial partners EG&G and Sverdrup Corporation as the SSCL contractor in January 1989.

In Wojcicki's words,

> The OER [Office of Energy Research] would have liked to manage the SSCL as a procurement where the overall integration responsibility for the project would rest with DOE headquarters. There would be two contractors—a scientific one, i.e., the URA, and a general industrial contractor (GIC) responsible for the construction of the Laboratory. This plan was presented to a group of senior reviewers on May 4–5, 1989, and their negative comments led to rejection of the scheme. (Wojcicki 2009, 273)

While the scheme may have been "rejected," Hunter's attempts to implement it anyway created havoc during the critical launch phase of the SSCL. I watched these events closely with growing distress until

THE SUPERCONDUCTING SUPER COLLIDER

Boyce McDaniel contacted me in desperation on behalf of the SSCL BOO, seeking my intervention at the highest possible level (see Wojcicki 2009, n. 44). I immediately drafted a letter to Knapp, URA's CEO and "our man in Washington." When they saw what I had written, Knapp, Panofsky, and Schwitters called to talk me out of sending it, as "inappropriate at this time." I was disappointed at what seemed to me a faintheartedness that would feed bureaucratic suspicions that URA would not provide "tough" management. I withheld the letter, but Knapp did arrange high-level meetings to press the issues. Readers may judge their seriousness from the following condensed and lightly edited version I wrote and retained in my personal files.

July 31, 1989
Dear Ed,

I am writing as Chairman of the URA Board of Trustees to express my deep concern regarding the impact of Department of Energy actions on the viability of the SSC project. . . . I would like you to take steps to communicate this concern to the . . . administration in a manner that it will be taken seriously. We have a major problem here that cannot be ignored any longer.

. . . It is evident that the SSCL performance has been outstanding . . . a judgment confirmed by the glowing report of an independent DOE Review Committee chaired by Edward Temple that was distributed on Friday.[23] . . . This success, however, has been achieved in the face of extraordinary difficulties created by DOE actions (and inactions).

---

[23]One of several "Temple Reports," this one, dated July 7–18, 1989, supported URA's concerns and urged immediate actions that would be necessary to satisfy DOE's demands. A subsequent, longer report is more widely known in the accelerator community as "*The* Temple Report," officially "Report of the DOE Office of Energy Research Review Committee on the Site-Specific Conceptual Design of the Superconducting Super Collider," DOE/ER 0463P, September 1990. A third very important report was commissioned by the Secretary of Energy, Admiral James D. Watkins, following a hearing before the House Science, Space and Technology Committee: *Report on the Superconducting Super Collider Cost and Schedule Baseline*, DOE/ER 0468P, January 1991. This is the report Robert Simon referred to in his April 1992 testimony to the Oversight and Investigations Subcommittee of the House Science Committee, quoted above.

The contrast between the highly professional and sophisticated operation of the URA/SSC group and the ineffectiveness of the DOE office created to provide agency oversight has become a national embarrassment . . . specifically

1. DOE announced the creation of a separate office in January to provide oversight of the SSC project. To date the office has not been staffed, with the following consequences:
   - the traditional source of high-energy physics management expertise within DOE has been insulated from the initial stages of the SSCL development. Responsibilities of the existing DOE staff are unclear because the transition to the new organization remains uncertain.
   - the unstaffed OSSC has been unable to respond in a timely fashion to critical program activities such as release of the geotechnical and site survey plans, the approval of leased space, and the procurement process for the architect-engineer/construction manager contract.
   - the creation of a new office reporting independently of other DOE HEP activities signals a functional separation of the SSC project from other physics that is unrealistic and potentially disastrous. The objective of the SSC is not to build an accelerator; it is to make new discoveries in physics. Its mission does not differ from previous DOE efforts in this field.
2. The DOE is not communicating with the SSCL regarding significant management actions in specific violation of Article 22(f)(1) of the URA/DOE contract. [examples are cited]
3. The DOE has proposed a level of staffing for the OSSC that is extremely large: 79 people in Washington and 146 people in Texas.
   - . . . needless duplication . . . signals an intent to micromanage . . . significantly alters the overall SSC budget, and draws out the schedule of activity . . .
4. DOE has taken arbitrary actions that have impeded the efforts of SSCL to carry out its responsibilities under the contract . . . most small, but adding up to significant frustration and delay.

THE SUPERCONDUCTING SUPER COLLIDER   67

- DOE has failed to approve the SSCL Director's choice for Deputy
  Director, with no substantive explanation.
- DOE has requested presentations and reviews requiring extensive
  preparation with little advance notice.
- DOE has failed to act in a timely fashion to permit the SSCL to
  acquire space in Texas, which is badly needed to adhere to project schedules . . . DOE is proceeding almost as if it were deliberately trying to sabotage the project. . . . The Trustees are increasingly impatient with the situation. . . .

In the meetings that did subsequently take place, President Bush's
new secretary of energy, Admiral James Watkins, intervened with helpful directives in September, and later that year, in Wojcicki's words,
"Hunter was forced to resign and Decker again assumed the position of
an acting director of the [Office of Energy Research]. This new situation
made it easier for Watkins to mold the formal organizational structure in
such a way as to give him much closer personal control over the SSCL.
Watkins had the DOE SSC ('on site') Project Director, Joseph Cipriano
(another Watkins colleague from his Navy days), report directly to him,
short-circuiting the normal lines of authority in the DOE" (Wojcicki
2009, 273). This assessment is literally true, but I think it is misleading
as to Watkins's intentions and his situation at the time. He had been
sworn in only on March 29, and his mission from President Bush was
very clearly to "clean up the mess in the weapons complex," which he immediately attacked with great energy, clarity of vision, and a clear and
sensitive understanding of organizational behavior. At this point "science organization" with respect to the SSC community was probably impossible for any outsider to understand, but Watkins listened and took
action that I regarded as favorable to URA, given the circumstances.

Perhaps I only need to mention that the Chernobyl reactor that caught
fire in April 1986 had design features (carbon moderator and incom-
plete containment) in common with two important reactors in DOE's
classified nuclear weapons production facilities. Department of Energy
secretary John S. Herrington asked the National Academy of Sciences

to study the safety of the weapons complex, and its report was highly critical of DOE's management of the complex. It explained that

> during the course of the study a number of developments affected our work. First, the operation of the weapons complex came under increasingly intense public scrutiny and criticism. News articles concerning the complex appeared almost daily, several congressional hearings were held, and a wealth of detailed commentary about the complex was offered by a variety of individuals and organizations, such as the General Accounting Office and the Advisory Committee on Nuclear Facility Safety. . . .
>
> Second, the national administration and the upper management of DOE changed in early 1989. The Secretary of Energy, James D. Watkins, publicly expressed his dismay at the past performance of the Department in managing the weapons complex and stated his intention to make substantial, if not radical, changes. (National Academy of Sciences 1989, vi)

The Second "Exogenous Factor"

Earlier I advertised two "exogenous" factors that emerged during the 1970s, and grew during the 1980s, to have a persistent and dominating effect on the DOE national laboratories. The first was the "inexorable" sequence of postwar federal administrative reforms that created increasing demands for accountability in government operations and in large publicly funded projects. The second swept dramatically into DOE with George H. W. Bush's appointment of Admiral James Watkins as his secretary of energy in 1989.

DOE watchers at the time might be forgiven for thinking of Watkins himself as an "exogenous factor." He joined the administration at President Bush's request to address the serious set of environmental problems in DOE whose visibility had grown alarmingly even before Chernobyl during the late years of the Cold War. These problems were known to the physics community but did not seem to have immediate significance for the SSC, which dominated the concerns of the particle phys-

THE SUPERCONDUCTING SUPER COLLIDER   69

ics community. Laboratories important to the SSC and its detectors appeared to be largely free of these problems, and the community was taken by surprise when the admiral chose to implement an environmental and safety "culture change" for *all* DOE operations, not just the weapons complex.

We needn't have been surprised. That decade of national transformation that included the management reforms that so irritated Robert Wilson and complicated the SSC procurement also saw the introduction of numerous environmental laws. In the 1970s alone we saw the Clean Air Act (1970); National Environmental Policy Act (1970), which also established the Environmental Protection Agency and the White House Council on Environmental Quality; Water Pollution Control Act amendments (1972), the Endangered Species Act (1973); the Safe Drinking Water Act (1974); the Resource Conservation and Recovery Act (1976); the Clean Water Act (1977); and the Superfund Act of 1980. This was the era when the nation woke up to the potential scale of environmental consequences of modern industrial economic expansion. Rachel Carson's *Silent Spring* appeared in 1962. DDT was banned as a pesticide in the United States in 1972. Pennsylvania's Three Mile Island nuclear reactor suffered its core meltdown in 1979. The hazards of nuclear weapons testing had been widely recognized since 1963 when atmospheric testing was abandoned by the United States, but weapons production and underground testing continued and increased in both the Carter and Reagan administrations. The nuclear production program was not subject to the many new laws and regulations that governed commercial reactors (regulated by the Nuclear Regulatory Commission) and other industries; it remained a secret program regulated by the DOE.

When Watkins discovered the extent of "the mess" in the weapons complex, he was appalled. At his confirmation hearing on February 22, 1989, he said, "They don't have any line management that makes any sense. No one knows what's going on. The system is antique, out-of-date and in desperate need of change. I intend to effect that change. . . . You'd be aghast at the lack of attention in administering policy. There is an urgent need to effect a significant change in its deeply embedded thirty-five-year culture." Watkins said most of the problems evolved from a program that put heavy emphasis on producing weapons-grade material without much regard to environment, safety, and health. "So now we are paying the price for this long-term cultural misdirection."

On June 6, 1989, during the time frame URA was pressing for high-level relief from DOE "actions and inactions" that imperiled the SSC project, "some eighty agents of the Federal Bureau of Investigation and the Environmental Protection Agency arrived at the Rocky Flats [nuclear weapons production facility] to carry out a search warrant filed in the U.S. District Court of Colorado. The warrant authorized the agents to search for evidence of alleged criminal violations of RCRA [Resource Conservation and Recovery Act] and the Clean Water Act." The agents arrived in helicopters, ensuring a media-worthy grand entrance. When I later learned of these events, I realized that the admiral had in fact been generous to URA to hear out our concerns and take some action to address them. But why did he have to impose new formal environmental and safety requirements on *all* the DOE contractors? Accelerator labs do not produce large quantities of radioactive materials and never seemed to have had the same potential for disaster as the facilities that actually manufactured bombs. The new formal management requirements for environmental health and safety (ES&H) resembled the performance-based management already strongly being forced on DOE management and operating contractors to assure complete transparency and accountability for public funds (Principle of Assurance). And these new requirements were here to stay. Watkins enforced them with rigorous inspections and "Tiger Teams" who scrutinized every laboratory for infractions and imposed penalties for violations. The program of ES&H culture change increased expenses and schedule times everywhere.

In all its applications, formal implementation of performance-based management adds to the expense and complexity of management, but in ES&H management it operates in a curiously different way that potentially reduces its cost and justifies its universal application. For the sake of concreteness, I will refer to ES&H management as "safety management"—acting so as to maximize safety. An intuitively obvious

THE SUPERCONDUCTING SUPER COLLIDER   71

and measurable performance outcome for safety management might be lost workdays due to on-the-job injuries. But who can tell whether our safety track record is not just due to luck? Depending on the job, we shouldn't expect to have many injuries, and luck could be playing a bigger role in our excellent performance than we realize. Lost work days is a "backward-looking" measure. *It provides no assurance of future performance!*

There *is* a forward-looking measure of safety performance, and that is the existence of a formal safety management system. It works like this: Each work element can be broken down into smaller elements— placing a piece of heavy equipment might entail moving it into place, lifting it, setting it down, and securing it. An experienced team of riggers, perhaps working with engineers, can identify a safe way of doing each step and create a training program appropriate to the operation. Before performing that step, a supervisor checks to see that everyone on the team has received the proper training and certifies readiness to work. If something goes wrong, it is recorded, reported, and analyzed, and a new procedure is written for the next time around. This sounds like a lot of work, but each step tends to be stable, so frequent revisions are not necessary. It is necessary, however, for workers to pledge their commitment to do the work according to the carefully developed plan. If so, then we can be assured that the work will be done as safely as we know how. That is the forward-looking principle of assurance embedded in formal safety management. It minimizes the possibility of luck in performance and exploits the continual improvement properties of the Plan–Do–Check–Act cycle. By contrast with its formal counterpart in cost and schedule management, the safety-related elements of a task are not subject to the exigencies of markets, labor costs, weather, and so on, that plague the application of performance-based management to larger pieces of the work breakdown structure. ES&H systems require lots of work to set up, but you don't depend on rapid change control or uncontrollable factors for them to function well. All workers have to do to is agree to take the training and do the work as planned. Supervisors have to pay attention to the outcome and have a way of changing the plan if there's a surprise.

After the Watkins reforms, all DOE labs were required to implement systems like this. The same basic structure works for all three elements of ES&H management. There is no reason not to capture its intrinsic value for all kinds of work, all the time. Watkins was wrestling with a large bureaucracy whose answer to poor performance was to force ever-increasing contractor compliance with a top-down conception of how work should be done. It was not until DOE itself learned how to do "integrated safety management" that it could deal effectively with the problems Watkins perceived immediately from his standpoint as an "outsider" with extensive management experience in a different but relevant environment.

References

Crease, Robert P. 1999. *Making Physics: A Biography of Brookhaven National Laboratory, 1946–1972*. Chicago: University of Chicago Press.

———. 2005. "Quenched! The ISABELLE Saga, Part 1." *Physics in Perspective* 7, September: 330–376; "Part 2." *Physics in Perspective* 7, December: 404–452.

———. 2008. "Recombinant Science: The Birth of the Relativistic Heavy Ion Collider (RHIC)." *Historical Studies in the Natural Sciences* 38: 535–568.

Forsythe, Dall W. 2000. *Performance Management Comes to Washington: A Progress Report on the Government Performance and Results Act*. Albany, NY: Rockefeller Institute of Government.

Hoddeson, Lillian, and Adrienne W. Kolb. 2000. "The Superconducting Super Collider's Frontier Outpost, 1983–88." Fermilab-Pub-01/076, August 2001. *Minerva* 38: 271–310.

Hoddeson, Lillian, Adrienne W. Kolb, and Catherine Westfall. 2008. *Fermilab: Physics, the Frontier, and Megascience*. Chicago: University of Chicago Press.

National Academy of Sciences. 1989. *The Nuclear Weapons Complex: Management for Health, Safety, and the Environment*. Washington, DC: National Academies Press.

Wheeler, John A., and Kenneth Ford. 1998. *Geons, Black Holes, and Quantum Foam: A Life in Physics*. New York: W. W. Norton.

Wojcicki, Stanley. 2008. "The Supercollider: The Pre-Texas Days: A Personal Recollection of Its Birth and Berkeley Years." *Reviews of Accelerator Science and Technology* 1: 259–302.

———. 2009. "The Supercollider: The Texas Days: A Personal Recollection of Its Short Life and Demise." *Reviews of Accelerator Science and Technology* 2: 265–301.

3

Managing a National Laboratory

In 1994, John Marburger stepped down as president of Stony Brook University and returned to research. Not for long. An episode that threatened the very existence of the nearby Brookhaven National Laboratory (BNL) soon called him back into active duty, facing him with new kinds of science policy issues.

BNL, located in Upton, New York, on Long Island, was founded in 1947 as one of the first three national labs, and the only one whose mission was basic research. It soon became a leading scientific institution worldwide, and its research has been awarded seven Nobel Prizes. In 1997, however, at the beginning of its fiftieth year, a small leak of slightly radioactive water from the spent-fuel pool of the lab's research reactor—a leak that did not affect drinking water and was not a health hazard—led to a media and political uproar. In its wake, the Department of Energy (DOE) Secretary Federico Peña fired BNL's contractor, Associated Universities, Inc.; BNL's director, Nicholas Samios, resigned (though Samios had earlier communicated his desire to step down); and activists called for the lab's permanent closure. In the fall of 1997 the DOE staged a competition for a new contractor. The competition for the $400 million contract to operate the lab was won by Brookhaven Science Associates (BSA), formed of a collaboration between Stony Brook University and Battelle Memorial Institute. BSA, which took over management of the laboratory in March of 1998, chose Marburger as the lab's new director. As BNL's new director, Marburger faced many new science policy challenges and new choices for ways to apportion his executive time.

In the first selection below, Marburger did what he had done in the course of leading the Shoreham Commission: he stepped back and painted a picture of the controversy in a way that took all concerns—those of the

74   SCIENCE POLICY UP CLOSE

scientists, the government, the community, and the activists—seriously. The second, third, and fourth selections show his attention to the press, to an individual, and to BNL employees. The fifth and sixth selections concern his response to another potentially devastating community fear: that the lab's new accelerator, the Relativistic Heavy Ion Collider, would create a black hole that would destroy the universe. In the final selection, on the tenth anniversary of his assuming the BNL directorship, he steps back once again to paint a picture of the lab's community relations on that decade.

—*RPC*

What Is the Future of Brookhaven National Laboratory?

New York Academy of Sciences
Science and Technology Policy Forum
February 3, 1998

In an address that he gave to the New York Academy of Sciences about a month before taking over the BNL directorship, Marburger does something very similar to what he did in the Shoreham Commission report—he retells the story of how the controversy arose in order to illustrate its causes, painting a big picture of the controversy in which everyone could recognize themselves. That big picture exhibits enormous respect for all sides, while acknowledging why the actions of each may be misunderstood or even regarded as misguided by the others. Near the beginning, in a remarkable flashback, he takes a frank look at the lab's vulnerability, calling what happened a "catastrophe" in the engineering sense, in which a system loses contact with its environment so that even a tiny perturbation can cause the system to fail or operate in a complexly different mode. Then Marburger makes an equally frank admonishment to those who thought that the fears generated by the tritium leak were overblown; mistakes that lead to small leaks can also lead to big leaks. He admonishes, as well, those who regarded the DOE's act of firing the lab management as overblown—who thought that the firing reflected "a brush too broad that was really meant for others"— meaning meant for weapons laboratories with far more serious environmental contamination issues. Marburger follows this with the observation that the BNL controversy was not simply a Long Island matter, but symptomatic of greater changes facing the country's scientific facilities in the

MANAGING A NATIONAL LABORATORY

post-Cold War era, changes that must be understood and addressed before an effective national science policy is possible. "In the absence of a national consensus on the basis for supporting science," Marburger says, "elected officials and news editors and agency administrators are listening to what the public wants. In the best cases, they try to make sense of the conflicting opinions and half-informed concepts current in the public idea marketplace. In the worst cases they propagate ignorance, zealotry and fear." Marburger then labels Secretary Peña's firing of BNL's contractor a "dramatic gesture" that was "intended to give a wake up call, not only to BNL but to the entire universe of DOE contractors, that their conduct had to conform to society's expectations for clean, safe operations that did not at all, ever, endanger the environment or public health." Ensuring such environmentally sound conduct is therefore an important lesson that needs to be learned by every large scientific institution in the twenty-first century.

—RPC

I have been asked to speak this morning to a question that has been asked many times during the past two decades, but from a continually changing perspective: What is the future of Brookhaven National Laboratory? Scientists began to ask this question after the cancellation of the intersecting beam accelerator (CBA), the ill-fated daughter of "ISABELLE." You will recall that this machine was canceled in the early 1980s when its aggressive magnet design took so long to mature that the all-powerful advisory committees judged that it could not compete with European efforts. The American high-energy physics community accepted this decision with the expectation that then-President Reagan would support the construction of a stupendous next-generation accelerator that our European colleagues could not hope to match. The Superconducting Super Collider was to be so large in scale it could not possibly fit on Long Island, and we all wondered what would happen to the vast circular trench that had been carved for ISABELLE into the gravel and pine scrub of Suffolk County.

Today the trench is being filled with RHIC, the Relativistic Heavy Ion Collider, about which more later, and the considerable talent of Brookhaven's high-energy physicists is being deployed in America's contribution to ATLAS, one of the great detectors that will extract information from the collisions of beams produced by Europe's LHC [Large Hadron Collider]. Europe succeeded after all by following through with its focused resources on a smaller, cheaper, faster version of the SSC. It is a scientifically riskier machine, because it does not cover the whole range of energies at which Higgs particles, the object of all this massive instrumentation, might be found. [In 2013, the Higgs particle was found at the LHC after an upgrade considerably boosted its power.—RPC]

During these years the question of Brookhaven's future was always a question of science strategy, of laboratory missions, of national needs for basic science and where they might best be satisfied. It was a question pondered by insiders, experts, commissions, and officials. It was not a question that Bill Smith, for example, a resident of Shelter Island and the founder and at times perhaps the only member of "Fish Unlimited," ever imagined he could influence. But in the spring of 1997 something quite unexpected happened, and suddenly the fate of Brookhaven National Laboratory seemed to many to hang so delicately in balance that even Mr. Smith might tip it over.

At that time I had no thought of seeking the director's post that Nick Samios had recently vacated. Nick had a brilliant science career that included the discovery of the omega-minus particle at Brookhaven's Alternating Gradient Synchrotron in the early 1960s. He had retired after a decade of careful maneuvering to assure BNL's continued dominance in basic physics through RHIC and a continually expanding set of lower-energy facilities, including the National Synchrotron Light Source (NSLS). Now it was time to celebrate the laboratory's fifty years of extraordinary accomplishment and to begin the systematic search for new leadership.

What actually happened, as you know, is that then-recently appointed secretary of energy, Federico Peña, directed the termination of DOE's contract with Associated Universities, Inc. to operate the lab and imposed interim management arrangements that were unprecedented in the history of the department. What had happened to turn the expected celebration into a trauma from which the laboratory will take years to

MANAGING A NATIONAL LABORATORY   77

recover? What does it mean for other labs? For American science in the closing years of the century?

As I grew nearer to the Brookhaven issue, still not suspecting how closely I would become involved, I asked myself these questions. I had discovered a delightful intellectual colleague in Robert P. Crease, a member of Stony Brook's excellent philosophy faculty. He is co-author with Charles C. Mann of *The Second Creation: Makers of the Revolution in Twentieth-Century Physics,* a commendable history of the Standard Model, and had just completed a history of BNL through the mid-seventies. We met for lunch every Tuesday and discussed the spectacle of the laboratory at bay as portrayed in the regional press. His insights came from years of studying the historical records of the laboratory in its regional context. Furthermore, he worked there several days per week and attended many of the community meetings that were held during that period to set forth the lab's view of its world.

On my side of our conversations, the insights came from my four-teen years of experience as president of Stony Brook and an occasional draftee to civic duty for Suffolk County or New York State, for example, as chair of Governor Cuomo's Fact Finding Commission on the Shore-ham nuclear power facility. I thought I understood what was happen-ing, but at the time I had no idea what one might do to alter what I saw as a tragic course of events.

After I was drafted in late summer to be the designated laboratory director for the team assembled by Stony Brook and Battelle Memorial Institute, I began to consider what I would say to the lab employees at our first meeting, were we to win the contract. About a month later I wrote the following introduction:

Imaginary Address to Employees of BNL
(written 1997)

The organization that brought me here today, Brookhaven Science Associates, came into existence as the result of a crisis in the laboratory precipitated by the unexpected termination of the contract between DOE and AUI [Associated Universities, Inc.]. What caused this crisis? Was it a random action, driven by politics and taken without regard to the value of the science performed here? I do not think so. I think it was a catastrophe in the modern technical sense—an abrupt adjustment of a system whose parameters had changed almost imperceptibly into a region of instability.

What has changed for Brookhaven are the conditions of the society in which it functions. In an odd sense, the very excellence of BNL's science insulated it from those changes and made it possible to continue to operate, at least for a while, as if they had not taken place. Society was willing to buy BNL's argument, up to a point, that good science is the bottom line and that the legalistic mechanisms of accountability being implemented elsewhere were an expensive luxury whose marginal benefit to society could not balance the reduction in scientific output necessary to create it here.

But the laboratory was operating unwittingly in a region of imbalance with its society. It was only a matter of time before a fluctuation would cause a catastrophic readjustment. The fluctuation that actually occurred was an unusual one, and large. If it had not occurred, the lab might have evolved by steady internal change slowly back into balance with its externals. Or it could have persisted in its metastable manner of operation and grown vulnerable to disruption by even smaller perturbations.

I want to stress my belief, having watched it happen, that the particular circumstances of this catastrophe are irrelevant. If we—and "we" now means you and me and our new BSA colleagues—adjust our response according to the specific details of what brought us here, we shall surely not move the laboratory into protective equilibrium. What we need to do is analyze carefully the conditions under which excellent research may be performed in the actual conditions of the society in which we find ourselves, and create those conditions here at the Brookhaven National Laboratory. Fortunately, and thanks to the thoughtfulness and energy of the laboratory leadership in place when this catastrophe occurred,

MANAGING A NATIONAL LABORATORY 79

much of this analysis has already been performed, and the first steps toward the necessary conditions have been taken. But much remains to be done.

I have referred to "legalistic mechanisms of accountability" as a symptom of societal change, and I used this ugly phrase to make a point. Many of our colleagues think that society wants us to do something unpleasant and unproductive, and that we should not capitulate. It might seem that somehow our laboratory has been swept with a brush too broad that was really meant for others. I do not think so. I think that society has reason to be concerned about the effects of the means by which it has succeeded in raising its standard of living even as its developing populations have grown manyfold.

Most industrialized nations have spoiled their environment and put their populations at risk. But the mechanisms of freedom that set America apart have produced a complex set of laws and practices that have at worst slowed the toxic side effects of progress and at best have induced a kind of harmony between the natural and the artificial. These laws and practices are restrictions imposed on business and governmental behavior—restrictions that continually grow tighter and that are continually taken more seriously. During the Cold War period, they touched only diffidently upon the fearful machinery that was thought necessary to maintain peaceful balance among superpowers. Society granted the machinery of nuclear and technological deterrence a kind of exemption from its censure. Some effects of this machinery, such as radioactive fallout from atmospheric nuclear testing, were too serious to be ignored. But much else was tolerated in the name of national security.

Brookhaven National Laboratory, too, although not a defense laboratory and doing little classified work, benefited from a national mood positive toward physics research. Although much was said during the Cold War about the benefits of basic research to society, the big appropriations for big science—the sequence of accelerators with ever higher energy at BNL, SLAC [Stanford's Linear Accelerator Center], Fermilab, and even ultimately the SSC—seem to have been made by a Congress that believed implicitly that such projects were important for national security. That, of course, all changed with the collapse of the Soviet Union and the end of the Cold War.

My imaginary talk to BNL employees broke off here. What I actually said to employees on that day in November when DOE announced that our team had won the contract was simpler but in the same spirit. I added more about what would happen next, but before proceeding, let me finish this line of thought.

Now Congress is struggling to assemble a new basis for federal science policy. The efforts of the leadership of the House Science Committee will provide fascinating material for future historians, much of it now available to us in real time on the World Wide Web. Science Committee vice chairman Vernon Ehlers has undertaken a project to (in the words of House Speaker Newt Gingrich) "develop a new, sensible, coherent long-range science and technology policy." Ehlers's recent editorial about this project in *Science* magazine asserts that "at the end of World War II, public support for funding of science was seen as critical in ensuring our nation's defense; the end of the Cold War has brought with it a vacuum in terms of a national imperative to justify research funding."

The important point I want to make this morning is that this absence of "national imperative to justify research funding" is a real phenomenon that was not simply invented by the House Science Committee. The discussion in Congress is accurately reflecting a public mood that has had real and impressive consequences for Brookhaven National Laboratory. What actually happened last year in eastern Long Island?

Ever since it took over the U.S. Army's World War I–era Camp Upton near Yaphank in 1947, BNL had been the target of criticisms about environmental contamination. In the 1950s complaints about the effects of radiation from the lab's new reactor (the Brookhaven Graphite Research Reactor, BGRR) began even before the uranium fuel was brought onto the site. Most of the other criticism was equally misguided, but some concern was justified. The lab sits over a sole-source aquifer from which Long Islanders draw their drinking water. As scientists

MANAGING A NATIONAL LABORATORY 81

learned more and more about the health and environmental impacts of pesticides, solvents, and weed killers, the Suffolk County legislature imposed increasingly tough constraints on the use of these chemicals. Some of them had been used at Camp Upton, and by 1996 it was known that disposal sites on the lab campus had created plumes of chemically contaminated groundwater that extended beyond its boundaries toward residential neighborhoods.

There are seven plumes in all under Brookhaven Lab, of which four have seeped beyond the borders. But they are deep—far below the depth from which drinking water is drawn by residences to the south, the direction of the groundwater flow. No well had water that violated federal standards. Nevertheless, by 1996 expressions of community concern had reached such a pitch that the Department of Energy agreed to hook up 1,300 residences in the area to the county water supply. Although homeowners must have felt somewhat relieved by this action, they also interpreted it as a sign that there really was something wrong with the water, and the net reaction to DOE's generosity was an escalation of fear and a demand for even more tests and monitoring activity.

It was in this context that a new underground plume was discovered, this one containing the radioactive element tritium, a form of hydrogen with two neutrons in its nucleus. Tritium is a weak beta emitter whose electrons are easily stopped by a sheet of paper, or by the bezel of a wristwatch whose fluorescent indicators are sometimes illuminated with its rays. The total amount of tritium in the new plume was less than that contained in a typical EXIT sign common in theaters. The plume had been there, undetected, for at least a decade. It had not flowed beyond the boundary and was unlikely to do so, given its twelve-year half-life. But it had flowed from a tank identified as a problem by Suffolk County health inspectors. Regional gas stations and industrial sites had long since replaced similar tanks with double-walled vessels to comply with county codes. BNL had not.

What particularly enraged the community was the fact of the leak and the long delay in detecting and reporting it. Many neighboring residents work at the lab or have relatives who work there, and they knew something about the hazards of radiation. County, state, and federal regulators assured the public there was no health hazard, and I think most people knew this and believed it. But for five decades they had been hearing nothing but reassurances from the lab that everything was safe, everything was under control. The tritium leak was a data point that contradicted all those assurances.

The leak was more than a data point. It was a red flag for environmental extremists, some of whom had been predicting disaster from the laboratory for decades. All the publicity the lab had been planning to celebrate its fiftieth anniversary in 1997 turned into media stories on the legacy of waste and hollow reassurances from BNL. Elected officials began to notice and to echo the community cries for increased accountability and improved management to prevent further such incidents. Their statements seemed to give credibility even to the most extreme critics, and the media genie was out of the box.

In response to the outcry, the Department of Energy conducted a review through its Office of Environment, Safety and Health. The review report appeared on April 1, 1997. On May 1, Secretary Peña came to the laboratory and announced that he was terminating the operating contract with AUI and taking five other steps to address problems identified in the report. In Peña's words, the action was "a result of unresponsiveness on the part of AUI to address DOE's needs and expectations for community relations and environment, safety and health stewardship." The arduous process required to identify a new contractor began immediately and was consummated in an unprecedentedly short period of six months.

As the laboratory scrutinized its facilities to discover other potential problems, it found an abandoned air cooling tunnel under the defunct graphite reactor (BGRR) into which water had leaked over the years, leaching traces of radioactive material that ultimately wound up in the groundwater, creating a second but smaller radioactive plume. This added further substance to the secretary's findings and further fuel to the media bonfire that flared throughout much of 1997.

At this point, the story could be unfolded in several directions, one toward the impact on the world of media perception exerted by the extreme environmental activists attracted to the wounded laboratory. An-

MANAGING A NATIONAL LABORATORY   83

other is the impact on science and scientists within the lab and else-where. Another is the changing culture of the Department of Energy itself. I want to talk specifically about what is happening at the labora-tory and what my company, Brookhaven Science Associates, is bringing to it.

A decade earlier BNL was not yet a "Super Fund" site, and the fate of the laboratory was decided by a relatively small number of people in Washington, DC, working together with panels of science experts. Poli-tics was certainly part of the national laboratory funding picture, but its role was always positive. No member of New York's congressional del-egation would think of opposing a significant operation within its labo-ratory. Today, even marginal community activists are able to get their views taken seriously. They are filling the policy vacuum that Con-gressman Ehlers described. In the absence of a national consensus on the basis for supporting science, elected officials and news editors and agency administrators are listening to what the public wants. In the best cases, they try to make sense of the conflicting opinions and half-informed concepts current in the public idea marketplace. In the worst cases they propagate ignorance, zealotry, and fear.

Far from being an arbitrary or thoughtless action, Secretary Peña's dramatic gesture was intended to give a wake-up call, not just to BNL but to the entire universe of DOE contractors, that their conduct had to conform to society's expectations for clean, safe operations that did not at all, ever, endanger the environment or public health. The question is, how to do this?

As I began to work with the Stony Brook and Battelle partners and the subcontractors they identified to help with the proposal, I learned that other private sector industries, and some national laboratories, had already mastered the kind of management that DOE wanted at BNL. The idea is to hold the science managers themselves responsible for all the critical operations that make their science possible. All BNL's facili-ties had environmental safety and health experts and overseers. But they owed their allegiance to an ES&H [Environment, Safety and Health] divi-sion, not to the science departments. The line managers responsible for getting the science done could easily view concern for the environment as someone else's responsibility. Although the analogy is too strong, I am reminded of the New Yorker's regard for litter: picking it up is Not My Job. With this separation of functions, an "us versus them" mentality can develop, and the idea that somehow these purely service functions of environmental safety and health are less important than the mainstream science activities and should suffer first when budget cuts occur.

What BSA will do at BNL is redefine management roles from the top down to make safety and concern for the environment part of the mainstream line management responsibility. We will redefine the relationship between science-oriented and service-oriented units to make the latter vendors to the former and to erase the idea that safety, for example, is a service task separate from science. It is a somewhat tedious business, requiring changes in organization charts, job descriptions, performance programs, self-evaluations, uniform procedures, increased communications, employee incentives, and so forth. But we know it works in other organizations where customers demand flawless, high-productivity performance.

The new way of doing things will make it easier for our neighbors, and DOE, to see what we are doing. It will be easier for us to document how we do it. And it will help us respond quickly to the vast number of inquiries from neighbors, activists, and what DOE calls "stakeholders," some quite far removed from the laboratory. Without this kind of internal organization and rethinking of how things get done, we will not be able to convince the public we can operate to its satisfaction. Consequently, we will not even have the chance to do the science we want and society needs. But in the absence of a war, hot or cold, I do not believe society will let us do otherwise. And why should it?

In my opinion the changes now taking place at Brookhaven National Laboratory are healthy. Society wants to see our science done, but under conditions in which we can guarantee benevolence to health and environment according to society's own standards as interpreted and rationalized by responsible government processes. If these conditions are not fulfilled, the traditional mechanisms for setting priorities for facilities, programs, and scientific directions will never even come into play.

MANAGING A NATIONAL LABORATORY

I am confident that the laboratory will rise to the occasion and enjoy another fifty years of excellent science.

Newsday Interview with John Marburger:
"It Takes More than Science to Do Science,"
Newsday "Viewpoints," December 30, 1997

Marburger's calm and attentive demeanor did much to reduce tensions and protect Brookhaven National Laboratory. But here he had a more complicated task than at Shoreham, for the burden of representation was more intricate: he had to represent the lab to the outside—to the community, media, and politicians. He also had to represent the lab to its inside, forging a mission for a multipurpose institution experiencing powerful centrifugal forces. He had to keep the respect of the scientists he managed even as he earned the trust of the community. Marburger therefore devoted some of his executive time to telling people how to "read" him, that is, how to appreciate the kind of task that he was facing. For instance, he would often mention to reporters the fact that, while at the University of Southern California, he spent three years building a harpsichord in his spare time, from a parts kit, to give him a diversion from the administrative problems of being a department chair. He also liked to point to his sports car, a 1970 MGB GT four-speed British coupe, that he maintained himself. Both seemed suitable symbols for Marburger the lab director, who had to keep hundreds of parts working together in finely tuned common cause. He also took care to make sure that the journalists read the symbols correctly. In November 1997, before he took over as BNL director, an article favorable to Marburger showed a picture of him with the car but, in a colorful but fanciful description, portrayed it as "spewing exhaust" and "shorting out in the rain." Marburger promptly fired back a letter, which *Newsday* did not publish, to correct this misapprehension. I don't ordinarily nitpick, it began, but that car "runs like a watch"; he added that it "doesn't smoke, has no electrical problems, and has plenty of power for its size and age." Given that the newspaper's story was favorable and the mistake innocent, why was he writing to correct such details? Because, Marburger continued, you never know when readers may draw incorrect conclusions—about his approach to safety once he takes over as laboratory director, for instance—from such details. "It is easy to make wrong assumptions about good information," Marburger finished.

"This time the story was cuter and no one suffered from the inaccuracies. Next time please be more careful."[1] Another example of caution about the appearance he projected appears throughout the following selection (slightly edited), a *Newsday* interview with Marburger shortly before he took over as BNL lab director. It provides an excellent example of how an administrator can speak to the press on delicate issues. At one point, for instance, Marburger is asked: "Do you believe in God?" (The sensitivities of that particular question had been prominently featured in the movie *Contact*, released just months before this interview.)[2] Marburger's response to this question, which engaged it straightforwardly but carefully, showed how he was able to navigate difficult circumstances with grace and respect.

—RPC

Q: Are you afraid of the Brookhaven lab?
A: I'm certainly not afraid of the Brookhaven lab. I think most people, because of the reporting that's been done on this issue over the past year, understand that Brookhaven is a scientific research laboratory. It does not make weapons. It does not have large amounts of radioactive material. It does not deal with large amounts of toxic chemicals. But I do think many neighbors are resentful of the laboratory for what they see as its too-casual attitude toward the environment.

Q: Is that a valid concern?
A: It is not acceptable to let a leak of radioactive material go on for ten years. I think the community is right to be concerned about that. The laboratory has not been sufficiently forthcoming about

---

[1]"To the Editors," by John H Marburger III, no date (but after Nov. 30, 1997), draft in the John H. Marburger III Collection, Box 14, "BNL 1997" folder, Stony Brook University Special Collections and University Archives. It was in response to Charlie Zehren, "He's Ready to Steer," *Newsday*, November 30, 1997, A6.

[2]In the movie *Contact*, the character played by Jodie Foster is interviewed by a commission to judge if she is fit for an interstellar journey. Thrown by the same question that Marburger was asked, Foster beats around the bush before replying reluctantly and defensively in the negative. The answer causes the commission to decide against her, although she later gains the position after a convenient plot twist.

MANAGING A NATIONAL LABORATORY   87

its operation. Many of the scientists there did not feel that the community understood them very well and didn't feel they were ever going to understand what kind of science they did.

Q: It used to be considered unpatriotic to criticize projects that were in the national interest.

A: That's just not there any more. We depend much more on public support for basic science.

Q: Especially because we don't see the payoff.

A: The payoff is very long range, and that's one of the arguments for government's supporting basic science. You simply won't find private industry justifying that kind of expense for something that's so far off.

Q: Obviously Brookhaven lab has become a politically hot issue.

A: That's right. It didn't have to develop into an issue. I'm sure that's what Energy Secretary Federico Peña thought as well. I believe the secretary values the lab and the science it does, and he's concerned that it wouldn't be able to continue to operate in the indefinite future if it didn't have political and community support.

Q: Can the lab really survive without that High Flux Beam Reactor, whose spent fuel rods were the source of the tritium plume?

A: The reactor is one of many components, many complementary facilities, the laboratory uses. It's not what makes the laboratory unique. What probably makes the laboratory unique is RHIC, the Relativistic Heavy Ion Collider. "Relativistic" refers to the speed with which the heavy ions collide with each other, close to the speed of light.

Q: Do the neighbors have anything to fear from a heavy ion collider getting out of control, sending ions through their living rooms or something?

A: No. Despite the fact it conducts very high energy, the current is extremely low.

Q: Is your mission to get this reactor back on line?

A: Not necessarily. My mission is to help Peña decide what to do with it. We need to convince ourselves we can run it without any more spills or accidents or danger to the environment. Is the department going to invest enough money to let us do that? The laboratory would not close if the reactor were to shut down.

Q: How did you react to Rep. Michael Forbes [R-Quogue] saying the reactor should be shut down?

A: Congressman Forbes is a politician and not a scientist. He is doing his job and I'm doing my job. If the laboratory were as concerned about Forbes's constituents as he is, he may not have taken the position he has taken.

Q: Is there anything you have learned in the study of physics that helps you run human institutions?

A: I know how things like reactors and accelerators work. I understand scientific issues, and I can help explain those to people.

Q: Are you religious? Do you believe in God?

A: I don't adhere to any particular organized religion. I believe there are mysteries in the universe that we don't understand yet, and perhaps never will. I grew up in the context of the Methodist Church, and it helped me a lot. I have nothing but affection for the church. The question of believing in God and supernatural forces is difficult for me to answer because I understand so much about physical forces in the universe and so little about human forces.

Q: Is using the collider almost like playing God, in the sense of trying to get to where matter and energy connect?

A: Oh, of course not! I don't think that playing God enters into this at all. RHIC is nothing more than a big microscope to help extend our senses and allow us to perceive things that we couldn't perceive with our own naked eye.

MANAGING A NATIONAL LABORATORY   89

Q: Last year a book came out by John Horgan called The End of Science. His premise is that all the major discoveries have been done and all that's left for scientists is to connect the dots. Do you believe that?

A: No, I don't think that's true at all. I think that we do not yet understand the basic structures of space and time and matter. We still don't know where mass comes from.

Q: As a physicist, how do you reconcile the legacy of physics, which means, of course, Hiroshima and Nagasaki? Does that weigh on you at all?

A: Those awful events were as much a consequence of electrical engineering and mechanical engineering. Many factors contributed to this disaster. [Note added after publication: This answer misrepresents my view. The decision to build and deploy nuclear weapons occurred in a complex context in which there were many actors, not just physicists, and that it is misleading to say that the legacy of nuclear weapons is "the legacy of physics." JHM]

Q: Is Brookhaven labs involved in weapons research?

A: Brookhaven laboratory is not a weapons laboratory. Brookhaven does a very small amount of classified work, primarily related to nuclear deterrence, perhaps what's called nuclear stockpile management. They tend to be paper studies.

Q: Do you read science fiction?

A: I don't read as much as I used to, but I did read a lot of science fiction as a child.

Q: Taking this administrative job is going to take you out of research for a while. How do you reconcile that?

A: I had to make that decision a long time ago when I got involved in administration. I enjoy administration. But it's true I miss doing science. I have a feeling of exhilaration when I'm working

on physics theories. One of the most rewarding things in my
life is to do physics. My first obligation is to the laboratory and to
try to solve its problems.

Q: *How does your style compare with your predecessor's?*
A: I plan to be in direct contact with the community, to be my own
spokesman rather than having somebody speak for me, and to
try to enable my senior administration and senior scientists to
do the same thing. I believe the next fifty years in Brookhaven
have to be ones in which the laboratory is much more out in the
public than it is now. We have the power to achieve discoveries
that would have been incomprehensible to previous generations.
That's what drives scientists at Brookhaven. They're not interested in weapons of war or poisoning people. They're interested
in discovering beautiful things about nature that everybody can
understand.

Q: *This new partnership with Battelle Memorial Institute of Ohio
sounds like kind of a shotgun marriage. You're scientists, and they're
a waste-management company. It seems like alien cultures coming
together.*
A: It takes more than just science to do science. The environment is simply too vulnerable on Long Island to expose to
casual behavior of people who have control of dangerous
materials.

Q: *[Scientists used to be] more cavalier.*
A: Yes, that seems to be the case. Scientists were more cavalier
before we knew what the consequences would be. So what's
necessary is to bring about a change in the way the laboratory
functions so everybody is more conscious of these things.
Everyone knows the tritium is nonhazardous, but at the same
time it shouldn't be there [underground], and you can set up
your work in such a way that you can prevent those [spills] from
happening.

MANAGING A NATIONAL LABORATORY   91

Q: How many hours of sleep do you get each night?

A: Five or six. If I'm interested in what I'm doing, I tend to wake up at the same time every morning, no matter what time I go to bed, and I'm usually reading or working till midnight.

Q: Is lack of sleep a prerequisite for being a scientist?

A: No, the prerequisite is being so interested that you are willing to sacrifice your time and comforts to do it. Real scientists are so interested in finding out things that other things become less important. In order to do science, there are a lot of things that don't intrude into our consciousness that are necessary to do science, which include staying healthy, having a reasonable family life, dealing with problems at home, talking to your neighbors about what you are doing, being careful not to do something that intrudes on somebody else's world so that they don't intervene in your ability to do science.

Letter to "Franklin" about the High Flux Beam Reactor
May 26, 1999

Activist opposition to Brookhaven National Laboratory focused on the High Flux Beam Reactor (HFBR), one of the key instruments at the lab. The HFBR had been taken off-line during the crisis that had precipitated Secretary Peña's firing of Associated Universities, Inc. as the lab's contractor and Marburger's taking over the lab directorship under Brookhaven Science Associates. Would the reactor be restarted? Secretary Peña took what looked to be a safe way to postpone the decision by ordering an environmental impact statement as a condition for the restart, and then he stepped down as DOE secretary, to be replaced by Bill Richardson. As the environmental impact statement was being prepared, through 1998 and into 1999, activists staged demonstrations against the reactor and circulated false claims of elevated cancer rates near the lab, and the lab received numerous letters that ranged from level-headed expressions of concern to ranting accusations. This selection is Marburger's response to one letter, written by a local middle school student named Franklin. As part of a classroom project, Franklin wrote, my friends "would like to know your beliefs, the ideas you have about this issue and how to solve it." In his reply, Marburger not only took Franklin's concerns seriously but laid out his thought processes in a thoughtful way. He addressed many issues touched on in Franklin's letter, and many that were not raised as well. The letter showed that Marburger was determined to take everyone's concerns seriously and prepared to conduct a genuine discussion with anyone who sought him out. As he will remark in another excerpt below, Marburger wanted people to know that "I thought their time was not worth less than my own." The penultimate paragraph of his letter to Franklin (beginning "Whom can we believe about these things? I will close by telling you how I decide") is an outstanding exhibit of Marburger's communicative skills. Marburger says that he is effectively in the same place as Franklin—confronted by a mixture of chaotic voices and the problem of which ones to listen to—and shares his counsel about how to reach an informed position. Marburger viewed the request to respond to a student not as a burden but as an opportunity, and his reply in effect is addressed to the entire community.

—RPC

Dear Franklin,

Thank you for asking about the environmental and health impacts of the High Flux Beam Reactor at Brookhaven National Laboratory. Many different issues are involved, and I have written briefly about some of them below. I can give short answers to some of the questions in your letter, but the answers are not what most people expect, so they need to be explained. For example, the topic "Whether the government should have closed down the High Flux Beam Reactor because of the radioactive contamination" has the easy answer "no" because there is no radioactive contamination from the reactor that poses any threat to health or the environment. People who have been reading the headlines would probably not believe such a statement, and yet that is the conclusion of all the government agencies that are charged with protecting public health and environmental quality. The HFBR was not closed for environmental or health reasons. You also mention saving energy as a possible issue, but the reactor consumes very small amounts of

MANAGING A NATIONAL LABORATORY   93

energy—just what is needed to run its pumps for cooling and for
air conditioning, heating, and lighting in its laboratory building.
And the work done with the HFBR does not have much to do with
radioactive chemicals or even nuclear physics, although elsewhere
in the laboratory we do make radioactive chemicals with cyclotrons
and other accelerators. So let me go systematically through the
issues as I see them, and perhaps from this set of short essays you
will get the flavor of how scientists view these things.

1. *Water.*   Long Island gets all its water from groundwater,
which sits in huge underground layers, called aquifers, beneath
the entire region. Rainfall washes down through the top layers of
soil and carries any contamination on the surface down into the
top aquifer, which is what most wells dip into. If your property
happens to be over a contaminated part of the aquifer, you need
to hook up to a public water supply whose wells suck water from
a clean part and pipe it to customers. Public water supply wells
are checked all the time for contamination, and they are shut
down if any appears. The water in the aquifers flows slowly underground, so once contamination gets into groundwater, it flows
with it, like smoke flowing with the wind away from a smoke
stack. So if there is a continuous source of contamination such
as an uncapped and unlined landfill, there will be a "plume" of
contamination flowing away from it underground. In our region,
the flow rate is about a foot per day. Roughly speaking, the
direction of flow in the top layer is away from the center of the
island toward the coasts. Where Brookhaven Lab is, the flow is
mostly from north to south. Unlike smoke in the air, contamination in underground plumes does not disperse very rapidly, so
the plumes remain more narrow than a smoke plume would.
Despite the fact that many such plumes exist on Long Island,
from old factories, gas stations, dry cleaning plants, and landfills,
the quality of drinking water is very high because the contamination is so localized. Fortunately, a single source cannot con-
taminate much of the aquifer. Because of the enormous volumes

of the aquifers, water is not scarce on Long Island. The best place
for more information on this topic is the Suffolk County Water
Authority.

2. *Nuclear reactors.*   "Reactors" are vessels where reactions
occur, so they can come in all sizes. Chemical reactors can be as
large as the huge plants we see along the Delaware River, or they
can be as small as a cooking pot. The same is true of nuclear
reactors. Like chemical reactors, nuclear reactors are designed
differently for different purposes. A power reactor is designed to
boil water to make steam to drive electrical generators, so it runs
hot and contains a huge amount of energy in its core. A research
reactor is designed to produce beams of neutrons to illuminate
small samples of materials under study. It runs cool, less than
boiling temperature, contains very little energy, and is typically
one hundred times smaller than a power reactor. Brookhaven's
High Flux Beam Reactor is a research reactor whose core is
about twice the size and a little warmer than a hot water heater
such as you would find in the basement of a home. The domed
structure you see in pictures of the HFBR is the entire building
housing the reactor itself plus all the laboratories, offices, conference rooms, shops, and so forth, that scientists and operators
need to do their work. See our website at www.bnl.gov for
pictures and more information about the HFBR.

3. *Nuclear reactions.*   All matter is made of atoms, which contain
very light electrons held near a nucleus by electrical forces. The
nucleus itself contains protons and neutrons ("nucleons") held
together by a force similar to electricity but stronger, called the
"strong force." Atoms can be drawn together by the electrical force
to make molecules. Chemical reactions occur when the atoms of
molecules are rearranged to make new molecules. Nuclear reactions occur when nucleons are rearranged to make new atomic
nuclei. The products of chemical or nuclear reactions sometimes, but not always, include heat and light, as in a gasoline

engine or a bomb. What you get out depends on what you put in and how the reactants are combined. Because the nuclear forces are much greater than electrical forces, the products of nuclear reactions usually come off with more energy than in chemical reactions. Where a burning candle (chemical reaction) will emit visible light, a "burning" nuclear reaction might emit gamma rays (light with very short wavelength), electrons ("beta" particles), or helium nuclei ("alpha" particles—two neutrons and two protons bound together).

4. How nuclear reactors work.   Larger nuclei, like uranium with more than two hundred nucleons, tend to break up ("fission") into smaller pieces when they are disturbed. The species ("isotope") of uranium with 235 nucleons breaks up when slow neutrons enter their nuclei. Among the broken pieces are two additional neutrons that can go on to cause other nuclei to fission if they are slowed down somehow. Heat is also released in the reaction (it's "exothermic"). Like exothermic chemical reactions, the heat can be released explosively or gradually. In the HFBR, an alloy containing uranium 235 is made into "fuel elements" that are immersed in an aluminum tank containing "heavy water." This is water whose hydrogen nuclei, which ordinarily are single protons, each have an additional neutron (hydrogen with one proton and one neutron is called "deuterium"). The reason for the heavy water is that it slows down neutrons very effectively, which gives them more time to interact with the uranium nuclei. If the neutrons go too fast, they don't do much to the nuclei. The water is also circulated to conduct the heat of fission away. In power reactors the heat is the important reaction product. In research reactors the neutrons are the important product. Control rods made of a material that absorbs slow neutrons are inserted among the fuel elements to stop or slow down the reaction. In the HFBR, hollow tubes are inserted into the fuel assembly to let some of the neutrons out into the experimental area where scientists place samples of material in their path and observe how the neutrons are deflected. When the materials are viewed in the "light" of the neutrons, scientists learn things about them they could not detect in ordinary light. Only the HFBR and one other smaller reactor in the United States were designed specifically to produce well-defined neutron beams. The best facility for this purpose in the world is at the Institut Laue-Langevin (ILL) in France. It was modeled after the HFBR, which then became the second best. Reactors do "age" as their neutrons displace atoms in the walls of the vessel that contains the cooling water and the metal becomes brittle. Embrittlement has proceeded very slowly at the HFBR, and even after thirty years of operation it appears to have many more years of useful life remaining.

5. *Nuclear fuel issues.* The fission products are much more radioactive than the original uranium. Fresh fuel is perfectly safe to transport and to work around. Spent fuel is too "hot" to handle because the fission fragments spit out high-energy particles as their nuclei rearrange themselves into more stable forms. At the HFBR, the spent fuel rods were lifted from the core vessel with a remote-control crane and deposited in a pool of water, the "spent fuel pool," to cool off and stabilize for a while. Then they were loaded into special metal boxes that shield the remaining radiation and trucked off to a storage site out of state. The HFBR fuel elements are small compared with power reactors and give off much less heat from the residual decay of their fission products. The fission products remain inside the solid fuel elements—they are not loose or detached in any way, but part of the solid alloy interior of the elements. Some opponents of nuclear reactors are concerned about the hazards of manufacture of the fresh fuel and disposal of the spent fuel. I do not believe these are serious issues for the HFBR because of the very small amounts, small sizes, and small heat generation of research reactor fuel components compared with power reactors. The HFBR fuel was manufactured from excess national supplies

MANAGING A NATIONAL LABORATORY   97

of enriched uranium whose production did not produce a need for additional hazardous mining or refining operations.

6. Emission of contaminants from the HFBR.   When operating, only trace amounts of the uranium or its fission fragments, except the neutrons, ever leave the solid fuel, which is sealed in metal tubes. Fragments that might escape through any imperfections in the tubes, and that can be dissolved in the cooling water, are easily removed by special filters. The neutrons, however, can enter the nuclei of the surrounding air, the cooling water, the walls of the vessel, and other mechanical components, and cause nuclear reactions within them. They can also simply knock the atoms out of place ("dislocate" them) in surrounding solid materials inside the reactor. Most of the new nuclei created in these reactions are unstable and decay quickly to stable forms. For example, neutrons can activate the element argon, which forms about 1% of air, to make it radioactive for a few hours. Some activated nuclei take longer to decay, and these remain "radioactive" long enough to be hazardous to workers when the reactor is shut down for maintenance. The only radioactive nuclei that live long enough to be dangerous, and are not tied down to the solid components of the reactor, are the altered nuclei of the hydrogen atoms in the cooling water. Recall that in the HFBR the cooling water is "heavy" so as to moderate the neutrons. Occasionally a neutron will attach itself to a heavy hydrogen nucleus, giving it two neutrons in addition to the original proton. The resulting isotope of hydrogen is called "tritium," which unlike deuterium is mildly radioactive. Although the tritium is confined to the primary cooling water system, normal operations occasionally expose this water to air, so some of it can evaporate and be drawn into the air conditioning intakes inside the building. The output of the air conditioning is vented to the environment through a stack whose height is designed to allow any evaporated tritium to disperse to undetectable levels before it can return to ground. Atmospheric radiation monitors located around the site operate continuously to warn of any elevated level of tritium in the exhaust. The air going up the stack is the same as the air that HFBR workers breathe. All workers are required to wear radiation monitors while in the building. The level of tritium is so low that it cannot be detected with these monitors. In my opinion, this source of radioactivity in normal operation is completely negligible, and has not caused a health threat to workers or to the public. Water vapor in the air can also condense in air handling equipment and on walls and surfaces, where eventually it can reach ordinary floor drains, so some tritium might be expected to show up in the water going to the sewage system. Monitors are placed to detect any radioactivity entering the sewage system, and valves are located to divert the fluids into holding tanks if undesirable levels are detected. The amounts of tritium in the drains are undetectably small unless there is a major accidental spill of primary cooling water into the drain system due to carelessness. The amounts of tritium in any case are extremely small, as is evident when you consider where it comes from.

7. The HFBR spent fuel pool leak. When the fuel assemblies containing spent fuel are lifted by the crane from the reactor vessel, they still have some cooling water clinging to them. They are allowed to drain and dry off, but when they are finally swung over to be lowered into the spent fuel pool, their surfaces still have traces of the tritium that was created by neutron activation of the heavy cooling water. So the spent fuel pool had a small amount of tritium in it. At the HFBR, this pool was built like a swimming pool, and it apparently had hairline cracks in its walls and floor, too fine to detect, through which water could slowly percolate to the ground beneath the floor. The Suffolk County Sanitary Code now requires such tanks to have a double liner to prevent leakage, but when the HFBR was built in 1968, the code was not in effect, and it was thought that leakage would be negligible. The leakage rate was less than or similar to the rate at which evaporation occurred from the top of the pool, so it was not noticed. Over a period of years, the leaked water accumulated and penetrated down to the water table, which is about fifty feet below the surface in this area. When it got to the groundwater, it flowed with it, creating a narrow plume in the direction of the groundwater flow. The plume was so narrow that it went past monitoring wells (that were not placed ideally downstream from the reactor building) and remained undetected for more than a decade. When new, better placed monitoring wells were constructed, they detected the plume through the radioactivity of the tritium. Laboratory management immediately notified the Department of Energy and other state and federal environmental agencies. At the time, early in 1997, these agencies were struggling with an entirely different problem of groundwater contamination from the lab. Since its beginning in 1947, the lab had disposed of its waste in a series of landfills that were established when the site was an Army base as early as World War I. From these, and from other sites such as vehicle repair locations, chemical substances like solvents and pesticides had leached into the groundwater and created plumes that had flowed beyond the laboratory's south border under the surrounding neighborhood. Eight years earlier the lab had been declared a federal "Superfund" cleanup site because of this and other contamination. The Department of Energy was angry at the discovery of the leak from the spent fuel pool because it had obviously been going on for a long time and would have been detected sooner had the lab's groundwater monitoring system been better. The department terminated its contract with the company that managed the lab and brought in a new contractor in March 1998 to improve the environmental management. I work for the new contractor, Brookhaven Science Associates.

8. Environmental and health impacts of the tritium plume.   Public health standards are set by governmental agencies such as the U.S. Environmental Protection Agency (EPA) and New York

State's Department of Health. To set a health standard for environmental contamination, you have to postulate a way for the contaminant to come into contact with people (a "pathway"). That will allow you to estimate how much of the contaminating substance gets into the body. Then you need to know the relation between the amount of the substance in the body and the occurrence of health effects. Many substances are harmless in small amounts, so the environmental standards usually state a maximum amount of substance that will be permitted to get into people's bodies. In the case of the tritium plume, the facts that it is confined completely on the laboratory property, is far from any drinking wells, and is so far underground that its radioactivity could not reach anyone, mean that there is no pathway and therefore no health threat. The total amount of tritium in the plume was very small, comparable to the amount used in an ordinary theater "exit" sign (the green ones use tritium). Some hydrogeologists have raised the possibility that eventually the plume could reach off-site, but that now appears not to be possible because the decay half-life of the tritium is considerably shorter than the time it will take the plume to reach the site boundary. As for environmental damage, there is no pathway to living things either in the present or in the future, so there is no issue of environmental or health impact at all in connection with the tritium plume.

9. *Threats from non-normal operation of the HFBR.* Most controversy surrounding nuclear power reactors centers on two issues: the fuel cycle from mining to disposal and storage, and the consequences of an accident such as that at the Three Mile Island plant in Pennsylvania or Chernobyl in the former Soviet Union. Since the HFBR is very much smaller than a power reactor, and is designed differently, both these issues are much less significant. I have already mentioned the fuel cycle above, and will here only talk about accidents. Since research reactors have very little energy in them, it is difficult to imagine an accident that could

MANAGING A NATIONAL LABORATORY 101

lead to health consequences to the public off-site. The most hazardous material in the reactor is the spent fuel in the metal of the fuel elements. A hazardous accident would need to melt or pulverize the metal and then spread it beyond the boundary of the laboratory, more than a mile away. I cannot conceive of anything that would do that, including direct hits by airplanes, catastrophic earthquakes, or meteor impacts. But to make sure, the engineers responsible for safety design just assumed that some kind of accident could take place that would throw the spent fuel into the air and break open the walls of the reactor building to allow the wind to blow fission fragments toward the community. Then they calculated the amount of radiation that would be detected at the site boundary under the worst conditions and compared it with standards of exposure set by federal health agencies. In what they call the "worst case credible accident," they discovered that the radiation levels would not exceed "protective action guidelines" above which residents would be advised to take some action (such as staying indoors) to prevent exposure to radiation. It is important to keep in mind that the hazard of radiation from a reactor is not in a blast where the radiation comes from a central point, but from particles carried by wind or water that eventually get inside your body and then decay radioactively, disrupting the molecules in the machinery of your cells. The radiation itself does not go very far. It is the particles that carry the radiation that have to be transported somehow to make a pathway. The most dangerous fission fragments do not dissolve in water and are not gases at normal temperatures, so they do not "float" but rather stick to surfaces and stay there. Research reactors are not dangerous to the public because they do not have enough energy to transport the hazardous material into public places.

10. *Social processes that led to the HFBR decision.* The Secretary of Energy decided not to restart the HFBR because keeping it in a state of readiness was expensive and was being paid for with

scarce funds meant to support scientific research. Public opposition, which translated to congressional opposition, made it difficult to predict when the reactor might once again be available for research. Scientific competition being intense, the secretary's science advisors recommended putting the money into facilities they judged to be more immediately productive. Although the HFBR was important for some kinds of science, and may even be unique, the Department of Energy decided that the loss had to be accepted in the face of other important science that competes for the same funding. Such arguments require a lot of information and a lot of judgment to make confidently, and therefore they were controversial. As to the public opposition that created the context for the decision, most members of the public never make their views known because they are not confident of their knowledge of the issues. Long Islanders care about the environment, and they fear for the quality of their drinking water. If they perceive that the HFBR is a threat to either, then they would oppose its operation. Public attitude is a matter of perception, and perception depends upon, among other things, where you get your information and whom you trust. It is among these difficult questions of public information and public confidence that one must search for the ultimate reasons for the loss of the HFBR. For my own part I believe the Secretary of Energy was correct when he said that there were no environmental reasons standing in the way of operating the HFBR.

11. *Factors affecting public attitudes toward the HFBR.*   The historian Spencer Weart has written a long book titled *Nuclear Fear* (Harvard University Press, 1988) that describes in detail how everything "nuclear" has become a negative symbol. Obviously people living near the Chernobyl reactor really did have something to fear, so it is foolish to discount the dangers of devices that employ nuclear reactions, even when they are not weapons. Public opinion is divided between those who would avoid any application of nuclear reactions because some applications have

MANAGING A NATIONAL LABORATORY   103

caused trouble, and those who view nuclear reactions as a useful
tool whose dangers can be avoided with reasonable precautions.
People in the second category make their case by pointing to
chemistry, which can create lethal substances, but which can be
controlled to produce useful substances safely. Today nuclear
reactions are better understood than chemical reactions (because
there are only two kinds of nucleons but about a hundred kinds
of atoms, and because electrical forces have much longer ranges
than nuclear forces), and it ought to be possible to design
devices and processes that use nuclear reactions safely. Some
people in the first category believe that the products of nuclear
reactions are always dangerous and that humans are too un-
trustworthy to act safely all the time, so all nuclear work should
be avoided. Between these positions are many people who
acknowledge that some applications are dangerous and some
are not, and the most dangerous ones should be avoided while
the less dangerous ones should be permitted. This seems the
most reasonable position to me.

12. How much radiation is safe?   What is radiation, after all? It is
whatever carries energy to us from its source. Scientists talk about
acoustical radiation, radiant heat, even seismic radiation. These
are low-energy forms of radiation that affect us through our
senses. Higher-energy forms of radiation, including the "alpha,"
"beta," and "gamma" emissions from radioactive nuclei, are rare
(but not zero) in normal environments, so we have not evolved
sensory responses to them (an exception might be suntan or skin
color for ultraviolet radiation from the sun). The radiation we
should worry about has enough energy to break apart the
molecules our cells need to perform the functions of life. This is
called "ionizing radiation" because it can break chemical bonds
and create loose ends that are electrically charged ("ions" are
charged atoms). It is important to understand, however, that there
are many mechanisms other than ionizing radiation that can
damage our cellular machinery. The very reactions of normal

cellular metabolism, oxidizing reactions, have side effects that continually break apart DNA, RNA, and other important cellular molecules. We know that cells have repair mechanisms that come into play either to fix the damage or to dismantle the damaged parts to prevent them from gumming up the works. Remember that biological molecules are complicated. DNA, for example, is a double helix with two strong "rails" winding around each other. Low-energy damage typically breaks one "rail" but leaves the other intact. Higher-energy processes can break both rails. Scientists, some of them at Brookhaven Laboratory, are still discovering repair mechanisms for different kinds of damage. Most scientists who study these things think that at the level of ionizing radiation that occurs in Earth's normal background, cell repair or scavenging mechanisms exist to prevent long-term damage to our bodies. My own attitude about the safety of radiation is that if it is within the normal range of background radiation on Earth, I do not believe it will have a harmful effect on humans or the environment. Not everyone agrees. Some people think that there is a small but finite probability that even one ionizing radiation event can lead to a fatal cancer. This is called the "no threshold" hypothesis. There is no observational evidence for this belief, and not enough detail is known about cancer or cell repair mechanisms at this time to estimate the probability of cancer at very low radiation levels. In the absence of knowledge, governmental health agencies have accepted the "no threshold" model and try to guess the probability at low levels from data at high levels of radiation. Scientific evidence is mounting, however, that the "no threshold" hypothesis is wrong, and I expect to see an increasingly sharp debate between scientists and health advocates on the issue.

Whom can we believe about these things? I will close by telling you how I decide. I tend to pay more attention to people who try to make distinctions between large and small effects than to

MANAGING A NATIONAL LABORATORY   105

people who make sweeping statements on either side. I listen to people who seem to be willing to change their minds based on new information. I am skeptical about the views of people who use dramatic or accusatory rhetoric to make their points. I tend not to listen to people who stereotype "scientists" or "bureaucrats," on the one hand, or "environmentalists" or "activists," on the other. I value humility in the face of ignorance, and courage when the facts are clear. Above all, I never give up trying to understand everything from my own knowledge and learning enough to make sound judgments without relying on the opinions of others. Science is made by ordinary men and women, and scientific knowledge is not reserved for some special class of people. I believe that if something is understandable at all, then all should be able to understand it.

Thanks for asking your questions. It gave me a chance to summarize the HFBR situation as I see it.

Sincerely,

John Marburger

Director

Responding to Closure of the HFBR

Statement after Hearing of the HFBR's Termination
November 18, 2000

The reactor's carefully planned restart process, involving creation of an environmental impact statement, was never completed. On November 16, 1999, DOE secretary William Richardson, who had met the month before with activists but had refused to meet with scientists involved with the reactor, announced the permanent closure of the HFBR. Lab officials and employees first learned of Richardson's decision from the news media. That snub added to their sense of outrage over Richardson's abortion of the deliberate process that the DOE had set up to help make the decision. Marburger was in a delicate position, having to announce the decision to the lab, express his disappointment, but also not antagonize Richardson and, indeed, figure out some way to minimize harm to the lab.

—RPC

Statement by Laboratory Director John Marburger regarding Secretary Richardson's decision not to restart the HFBR, issued September 18, 2000:

I am deeply disappointed by the secretary's decision not to restart the HFBR, one of the most productive research tools at Brookhaven National Laboratory. I understand that this decision was not made because of any conclusions about the environmental safety of the reactor, but because of the cost of restarting it. The cost to national science effectiveness is likely to be far greater.

The loss of this valuable facility will come as a blow to many who had invested their lives in research it made possible. Plans to preserve these investments must move ahead, and, in this connection, I am pleased that the secretary has expressed his support for the future of the laboratory. Secretary Richardson also expressed his concern for the employees whose jobs in connection with the HFBR are now in jeopardy. I share his concern and expect to work with the Department of Energy to assist them in developing their future career plans.

The untimely demise of this powerful research facility leaves a hole in the spectrum of capabilities for which Brookhaven National Laboratory has long been renowned. It will take strong support from those who have a stake in the laboratory's future—including the Long Island community, elected officials, and researchers—to fill this gap swiftly. I am grateful to the many friends of the laboratory who have already expressed their support. We will need all their help to assure the future excellence of this great laboratory.

Despite Marburger's statement, BNL's deputy director for operations, Tom Sheridan, publicly complained about Richardson's action in deciding to abort the restart process without input from scientists. Richardson strongly

MANAGING A NATIONAL LABORATORY   107

rebuked Sheridan and urged Marburger to fire him. Marburger interceded with Richardson to save Sheridan's job, for which Sheridan thanked him with a bottle of rare Irish single-malt whiskey.

—RPC

Disaster Scenarios I: Report on Fears That a Black Hole Will Destroy the Universe

Statement on the Report of the Special Committee
on RHIC "Disaster Scenarios"
September 30, 1999

Meanwhile, Marburger had to address media-fed fears that Brookhaven's new accelerator, the Relativistic Heavy Ion Collider (RHIC), would create black holes that would destroy the universe. The episode began in March 1999, when *Scientific American* ran an article about RHIC whose title, "A Little Big Bang," referred to the machine's ambition to study forms of matter that existed in the very early universe. A letter in response, from the founder of a botanical garden in Hawaii, asked whether scientists knew "for certain" that RHIC would not create a black hole. *Scientific American* printed the letter in its July issue along with a response by Frank Wilczek of the Institute for Advanced Study in Princeton. Wilczek called RHIC's ability to create black holes and other such doomsday ideas "incredible scenarios." Amazingly, however, he then went on to mention other doomsday scenarios such as new forms of matter called "strangelets," saying that they were "not plausible" but had a tiny fraction of probability. This provoked the *Sunday Times* of London, on July 18, 1999, to carry an alarming article: "Big Bang Machine Could Destroy Earth." The article claimed that RHIC, then nearing completion, might create black holes or strangelets, which would either blow up the planet or suck it into oblivion. A caption to the story asked: "The final experiment?" Marburger, wanting to create a "speed bump" that would slow the hysteria, appointed a committee of eminent physicists, including Wilczek, to evaluate the possibility that RHIC could cause a doomsday scenario. The resulting report was so dry and technical that some felt it would do little to reassure those who were in fear of the collider. Thus, before it was put up on Brookhaven's web page, Marburger wrote a summary in non-technical language. He placed it at the beginning of the report, so that one

108   SCIENCE POLICY UP CLOSE

had to scroll through his summary first. It is a sterling example of how to help the public predigest a technical report.

—RPC

In July 1999 I appointed an expert committee to assemble material relating to the safety of RHIC experiments. The committee, composed of Robert Jaffe (chair) and Wit Busza of MIT, Jack Sandweiss of Yale University, and Frank Wilczek of the Institute for Advanced Study at Princeton, has submitted its report. The report summarizes technical arguments that conclude there is no danger of a "disaster" at RHIC. Because the language of the report includes technical terms and concepts that may be unfamiliar to many interested readers, I have summarized the contents below in a less technical form. Details may be found in the full report that is available on the Brookhaven National Laboratory website: www.bnl.gov. Material in quotations is excerpted from the report.

RHIC, the Relativistic Heavy Ion Collider, is a pair of circular particle accelerators located at Brookhaven National Laboratory that will accelerate the nuclei of gold atoms in opposite directions to nearly the speed of light. The energetic nuclei will be directed toward each other to form head-on collisions within experimental observation instruments at four stations around the ring. Digital "pictures" of the collisions will be analyzed with extensive computation to infer the behavior of matter at the instant of collision. The experimenters hope to see evidence of a form of matter called the "quark-gluon plasma," a state of matter that is thought to have existed throughout the entire universe a few microseconds after its creation in the Big Bang. Some people have expressed concern that the RHIC collisions could trigger a complicated process that would have disastrous consequences.

Three different kinds of "disaster" scenarios have been discussed in connection with high-energy particle collisions:

A. Creation of a black hole that would "eat" ordinary matter
B. Initiation of a transition to a new, more stable universe
C. Formation of a "strangelet" that would convert ordinary matter to a new form

MANAGING A NATIONAL LABORATORY   109

In connection with these scenarios, the authors of the report "have reviewed earlier scientific literature as well as recent correspondence about these questions, discussed the scientific issues among ourselves and with knowledgeable colleagues, undertaken additional calculations where necessary, and evaluated the risk posed by these processes. Our conclusion is that the candidate mechanisms for catastrophic scenarios at RHIC are firmly excluded by existing empirical evidence, compelling theoretical arguments, or both. Accordingly, we see no reason to delay the commissioning of RHIC on their account."

The authors state that "issues A and B are generic concerns that have been raised . . . each time a new facility opens up a new high energy frontier. . . . There are simple and convincing arguments that neither poses any significant threat." Issue C is a new concern raised specifically with respect to RHIC collisions, and the report gives extensive information on this scenario.

A. Black holes.   A "black hole" is a spherical concentration of matter so great that its gravity causes the escape velocity at the surface of the sphere to equal or exceed the speed of light. Since light cannot escape, the sphere would appear black. Large black holes attract matter to themselves and effectively "consume" the material in their vicinity. Very small black holes evaporate quickly by a quantum mechanical mechanism that essentially overcomes the escape velocity limitation on radiation. Extremely small amounts of matter such as the two gold nuclei in a RHIC collision must be compressed to exceptionally small sizes to form the concentration necessary to create a black hole.

Appendix A of the report uses a well-known formula for escape velocity to assess the possibility of black hole formation. If all the mass available were compressed to the size of a single proton or neutron (of which there are 197 in each gold nucleus), the escape velocity would be eleven orders of magnitude (powers of ten) less than the speed of light. No mechanism exists to compress the matter even to this size. The report states that "collisions at RHIC are expected to be *less* effective at raising the density . . . than at lower energies where the 'stopping power' is greater." In other accelerators operating at both larger and smaller energies, say the authors, "in no case has any phenomenon suggestive of

110   SCIENCE POLICY UP CLOSE

gravitational clumping, let alone gravitational collapse or the production of a singularity [i.e., a black hole], been observed."

B. Vacuum instability.   This is an exotic possibility of which the report states that "physicists have grown quite accustomed to the idea that empty space—what we ordinarily call 'vacuum'—is in reality a highly structured medium, that can exist in various states or phases, roughly analogous to the liquid or solid phases of water. . . . Although certainly nothing in our existing knowledge of the laws of Nature demands it, several physicists have speculated on the possibility that our contemporary 'vacuum' is only metastable, and that a sufficiently violent disturbance might trigger its decay into something quite different. A transition of this kind would propagate outward from its source throughout the universe at the speed of light, and would be catastrophic."

The vast stretches of interstellar space are penetrated continually by swarms of energetic particles called "cosmic rays." Before particle accelerators, scientists studied high-energy particle phenomena by examining the tracks of cosmic rays in special detectors. To date, no accelerator, including RHIC, has been able to produce collisions so energetic that they could not be found in cosmic rays.

Stating that "cosmic rays have been colliding throughout the history of the universe, and if [a catastrophic] transition were possible it would have been triggered long ago," the authors cite a relevant 1983 study by physicists P. Hut and M. J. Rees. That study used data on cosmic ray properties that have since been updated, and the RHIC committee reexamined the issue using more modern data. Appendix B of the report gives technical details on this issue. Their conclusion is that the work of Hut and Rees remains valid and that "we can rest assured that RHIC will not drive a transition from our vacuum to another."

C. Strangelets.   Of this disaster scenario, the authors say that "theorists have speculated that a form of quark matter, known as 'strange matter' because it contains many strange quarks, might be more stable than ordinary nuclei. Hypothetical small lumps of strange matter, having atomic masses comparable to ordinary nuclei, have been dubbed

MANAGING A NATIONAL LABORATORY   111

'strangelets.' Strange matter may exist in the cores of neutron stars,
where it is stabilized by intense pressure. A primer on the properties of
strange matter . . . is contained in Appendix C."

Appendix C gives a great deal of technical information that will be of
interest to those who wish to know more about the kind of physics RHIC
is designed to explore. In this brief summary, I will give only some additional very simple information about the composition of matter, and
summarize the argument, again based on cosmic ray data, that reassures
us that RHIC collisions will not lead to a disaster through strangelet
formation.

The best theory of matter we have today, called the Standard Model,
explains essentially all experimental observations on matter to date at
scales smaller than atoms. This model regards all other small-scale
objects as composed of families of particles called *quarks* and *leptons*
and the forces that bind them together that have their own particles
called *gauge bosons*. There are six quarks with the somewhat whimsical
names *up, down, strange, charm, bottom,* and *top*. Only up and down
quarks occur in ordinary matter. Protons have two ups and one down,
and neutrons have one up and two downs. Atoms have nuclei made of
protons and neutrons in which nearly all the mass resides, and a number of electrons—which are leptons—moving about the nucleus at distances up to hundreds of thousands of times the radius of the nucleus.

All particles ever observed to contain "strange" quarks have been
found to be unstable, but it is conceivable that under some conditions
stable strangelets could exist. If such a particle were also negatively
charged, it would be captured by an ordinary nucleus as if it were a
heavy electron. Being heavier, it would move closer to the nucleus than
an electron and eventually fuse with the nucleus, converting some of the
up and down quarks in its protons and neutrons, releasing energy, and
ending up as a larger strangelet. If the new strangelet were negatively
charged, the process could go on forever. This is a simplified picture of
the strangelet disaster scenario.

The report presents arguments that suggest that the conditions for
this scenario will not be satisfied in RHIC collisions. In particular, say
the authors,

1. At present, despite vigorous searches, there is no evidence whatsoever for stable strange matter anywhere in the universe.

2. On rather general grounds, theory suggests that strange matter becomes unstable in small lumps due to surface effects. Strangelets small enough to be produced in heavy ion collisions are not expected to be stable enough to be dangerous.

3. Theory suggests that heavy ion collisions . . . are not a good place to produce strangelets. Furthermore, it suggests that the production probability is lower at RHIC than at lower energy heavy ion facilities like the AGS [Alternating Gradient Synchrotron accelerator at BNL] and CERN. . . .

4. It is overwhelmingly likely that the most stable configuration of strange matter has positive electric charge.

"However," the authors state, "one need not assess the risk based on theoretical considerations alone. We have considered the implications of natural 'experiments' elsewhere in the universe, where cosmic ray induced heavy ion collisions have been occurring for a long time." Theoretical considerations indicate that strangelet formation would be even greater for collisions of iron, of which some cosmic rays are composed, than for gold, which will be used in RHIC. The total number of collisions that will occur in RHIC over ten years turns out to be far fewer than the number of potentially "dangerous" iron-iron collisions that occur on the surface of the moon in a single day. For every production of a dangerous strangelet at RHIC, one expects one hundred thousand trillion to have been produced on the moon during its lifetime, any one of which would have converted the moon explosively to strange matter—a phenomenon that is known not to have occurred.

During the preparation of the report for RHIC, the authors were in contact with a European group of physicists studying similar disaster scenarios (A. Dar, A. De Rujula, and U. Heinz). This group made "worst-case" assumptions about what kind of cosmic ray collisions would produce dangerous strangelets in order to achieve the greatest possible assurance that they would not occur in RHIC. The report says of the European study that "they assume that strangelets are produced only in

MANAGING A NATIONAL LABORATORY   113

gold-gold collisions, only at or above RHIC energies, and only at rest in the center of mass. Under these conditions . . . it is necessary to consider ion-ion collisions in interstellar space, where strangelets produced at rest with respect to the galaxy would be swept up into stars. Dangerous strangelets would trigger the conversion of their host stars into strange matter, an event that would resemble a supernova. The present rate of supernovae—a few per millennium per galaxy—rules out even this worst case scenario." Thus, conclude the authors, "we demonstrate that cosmic ray collisions provide ample reassurance that we are safe from a strangelet initiated catastrophe at RHIC."

Readers wishing to learn more about RHIC physics will find a wealth of detail on the RHIC website at www.rhic.bnl.gov.

Disaster Scenarios II:
"What Is the Risk of Boating on Loch Ness?"

February 1, 2000

The media loved the "accelerator Armageddon" story—newspapers, magazines, and broadcast media recycled old pictures of nuclear destruction to accompany stories about Brookhaven—and Marburger struggled with a way to explain clearly to nonscientists why there was no danger. This is one unpublished item in which he worked out a clear and accessible way—a model of science communication. It consists of a thought experiment involving a supposed peril already familiar to the public that effectively makes the difference between real and hypothetical probability. It is a wonderful example of science communication about a volatile subject, and its form should be copied more frequently.

—*RPC*

Some responsible people believe Loch Ness harbors a creature of great size and potential danger to unwary boaters. Scientific arguments cast doubt that such a creature exists, but no fundamental laws of nature would be violated if it did. Extensive and systematic searches by reliable observers have not produced any evidence of the monster and have excluded all obvious possibilities for its habits and manner of survival. In view of the failure, however, of theory and experiment to definitively exclude a Loch Ness monster, it would seem prudent to estimate the risk to boaters.

In the spirit of cost-benefit analysis, it is necessary to develop a measure of risk that can be compared with gains. Gains might include the tourist trade. Against these must be set the risk of injury or loss of life through a monstrous encounter.

To derive the risk, one needs to know the probability that a creature exists with characteristics that would present danger to humans. Given a particular set of physical and behavioral parameters, one can work logically through trees of events to arrive at the probability of particular accidents or injurious incidents. But the trees have to start somewhere.

What is the probability of a phenomenon that has not been observed but is not definitely excluded by physical law? One might imagine a sequence of events consistent with known science, each with a probability of its own, that might be concatenated to produce a Loch Ness monster. Then the probabilities could be multiplied to give a final estimate. This approach has the virtue that it automatically produces a set of characteristics that can inform subsequent risk scenarios. It has the drawback that it depends very much on the imagination of the analyst. It is conceivable that no sequence may be found that produces a monster. Certainly the resulting hypothetical creature is not unique.

If it is not possible to deduce a probability through a chain of possible events, the concept of probability does not help us, and we may as well simply assume that the monster exists, just to get the analysis started. Assuming also the monster's characteristics then enables us to evaluate scenarios based on weather, likelihood of human activity coinciding with the creature's own active hours, and so forth. Then each scenario leading to injury could be analyzed further to identify safety measures that might reduce the probability of harm. If these mitigating precautions are taken, then for each set of assumed characteristics, the probability of harm can be estimated given the initial probability that the creature exists.

MANAGING A NATIONAL LABORATORY

This approach has the virtue that it can be carried out, but it mistakenly implies risk if the hypothesized creature does not exist. If it does exist, the risk estimate is accurate only if the analyst has correctly guessed its characteristics. Policy makers can optimize a safe pattern of behavior for any single set of assumptions, but the analyst can always change his or her hypotheses to produce a creature for which the mitigating policy is ineffective.

Probabilities of occurrence appear to have significance only in two cases: (1) phenomena for which the physical basis is completely understood, such as the random motion of molecules in a gas, and (2) phenomena that have been observed to recur over a period of time. In the first case, one can make estimates, based upon random initial conditions, regarding the probable behavior of the system. In the second, one expresses the probability as a frequency of occurrence. If the phenomenon is so rare as not to have occurred at all, and if we have no physical theory that predicts its frequency, then no quantitative estimate is meaningful.

Does it not require courage to embark upon Loch Ness? What is the source of this dread? Not reason, for it cannot guide us here.

Effective Risk Communication

Community Advisory Council 10th Anniversary
Brookhaven National Laboratory
September 18, 2008

In 2001, Marburger received a call from Washington sounding him out as to whether he might be interested in becoming science advisor to President George W. Bush. Marburger said that he would think about it, but his instinct was to decline because he liked the lab directorship and its responsibilities. Many scientists, however—both in Washington and in New York—strongly urged him to take the post. He accepted, was confirmed by the Senate, and arrived in Washington in the fall of 2001. In 2008, on the tenth anniversary of the changeover of the lab's directorship, he was invited back to address the Community Advisory Council (which had been created as part of the changeover), a key feature in the lab's new commitment to engage in a dialogue with the community about the risks posed by lab operations. He used the occasion, once again, to paint a "big picture" in which everyone felt at home, and made suggestions for improving lab-community relations.

—RPC

I want to thank Jeanne D'Ascoli, Marge Lynch and Sam Aronson [Marburger's successor as BNL director] for inviting me back to celebrate CAC's ten years of operation. Looking back, the intervening ten years seem very short. What seemed long were the meetings! I attended nearly every one because I knew they were important for the future of the laboratory, and I wanted the members to know I thought their time was not worth less than my own. So I kept myself awake with coke and chocolate chip cookies and forced myself to listen and tried to sort out how we could rebuild a sense of partnership with the community. As far as I can tell, that goal has been achieved. If you tell me all tensions between the lab and the community have disappeared, I would be very skeptical. But fortunately that's not necessary to have a constructive partnership.

Three years ago I gave a talk at a symposium inaugurating a Center for Risk Science and Communication at the University of Michigan in Ann Arbor. Of course I referred to my experience at Brookhaven, and tonight I'll tell you some of what I said on that occasion.

Among other things, I criticized some very smart people for misapplying formal risk analysis to the proposition that RHIC would create a dangerous new particle that would destroy the earth. That history is repeating itself now as the LHC starts up at CERN. If these experts could make such a mistake, I asked, how do we expect the general public to understand measures of risk?

Well, I said, *the general public is not seeking to understand measures of risk.* In my experience with the hypothetical dangerous particles at RHIC, and the not-so-hypothetical environmental contamination at the laboratory, the public did not care much about formal risk assessments at all.

In 1997 neighbors of the lab were outraged that radiological material—tritium, to be precise—had leaked from a research reactor (actually its spent fuel pool) into the groundwater. The lab scientists knew there was no danger for a variety of technical reasons, so in public meetings they attempted to give careful technical explanations of the relative risk created by the leak. You can guess how successful that strategy was. It's not that scientists reason so differently from nonscientists, but there were different interests involved, and a great deal of suspicion about motives. Basically, the scientists wanted to continue to do their work—which was in fact very beneficial to society—and the neighbors wanted to be protected from the hazards of radiation they perceived, no matter how important the science was.

From the scientists' perspective, the neighbors were demonstrably wrong. From the neighbors' perspective, the scientists had shown themselves to be irresponsible and unreliable, and they felt nothing the scientists said could be trusted. The scientists thought the neighbors' concerns resulted from inadequate understanding of science. The neighbors thought the scientists were arrogant and dismissive of the health impacts they perceived themselves and their families to be vulnerable to because of the lab's presence in their midst. The scientists imagined the solution in the long run to be better science education. The neighbors imagined the solution to be greater public accountability for the conduct of science by governmental organizations over which the public had influence through democratic political processes. The scientists viewed the neighbors' solution as unwarranted intrusion of politics and bureaucracy into the conduct of science. The neighbors viewed the scientists' solution as insulting.

I am oversimplifying, of course, but in my experience these attitudes are almost universal when the public perceives itself to be at risk from the actions of an establishment. The immediate issue is not science or education or even the facts about risks and hazards. The immediate issue is a mutual suspicion and loss of confidence in the good faith of the other side. "Risk communication" refers to the actions the organization takes to address this loss of confidence. The burden is on the *organization* to discover not only risk but whether a perception of risk exists, and to initiate contact with the community.

Organizations that understand risk communication are aware and in control of risk-producing operations and their potential impacts. They are prepared to demonstrate management responsibility for these operations from the top down. They understand and interact with the potentially affected community. They analyze threat scenarios, plan responses, and rehearse their response capability. Risk minimization is part of their institutional culture, and they take pains to demonstrate this to their publics.

When an incident occurs, successful risk communication requires immediate disclosure of incidents accompanied by all reliable information. Assurances and cautions must be backed by knowledge, and the information should be delivered in a neutral, objective, and serious manner. Successful risk communication requires extensive planning and institutional awareness so information about threats, risks, and responsive actions are ready, and trained personnel are ready to convey them. All this requires budgets and people.

Above all, the organization must take seriously the grand statements it makes about its commitment to operational excellence, public and environmental health and safety, and open and frank dialogue with the public, and this commitment should be expressed by the highest level of management.

For a decade now the Community Advisory Council has been crucial to this dialogue. Tonight it is my privilege to be able to thank you, not on behalf of the laboratory, but on behalf of the nation it serves so ably. From my perspective your service on this council has been a national service, and it has made a lasting contribution to the strength of our country. Thank you.

4

Presidential Science Advisor I

Advice and Advocacy in Washington

In June 2001, President George W. Bush, who had assumed office in January, nominated John Marburger to be the presidential science advisor, in which capacity he would also serve as director of the Office of Science and Technology Policy (OSTP). According to its web page (http://www.whitehouse.gov/administration/eop/ostp/about, accessed June 9, 2014), the mission of the OSTP, a White House office of about fifty people, is "to provide the President and his senior staff with accurate, relevant, and timely scientific and technical advice on all matters of consequence; second, to ensure that the policies of the Executive Branch are informed by sound science; and third, to ensure that the scientific and technical work of the Executive Branch is properly coordinated so as to provide the greatest benefit to society." Marburger was approved in October 2001, a few weeks after the terrorist attack of 9/11.

Marburger continued to serve as presidential science advisor and OSTP director for over seven years (October 23, 2001, to January 20, 2009), becoming the longest-serving U.S. science advisor since President Eisenhower appointed the first full-time science advisor in 1957. Marburger's experiences at Stony Brook University and then at Brookhaven National Laboratory prepared him for the post by giving him an appreciation for the need to prioritize, for the necessity to cultivate the often-overlooked "working parts" that make the system go, and for the necessity to be unflappable in the face of criticism. In Washington, the criticism was severe. While Marburger's critics on Long Island were largely local and intermittent, critics in Washington were nationally organized and orchestrated and a permanent feature of political life.

As both presidential science advisor and OSTP director he therefore wore two quite different hats. Most of his publicly debated activities relate to his role as presidential science advisor. Yet the OSTP was a creation of Congress, and its head required congressional approval. In that role he also served as an adviser to Congress—either to individuals or to groups such as senate or congressional committees that sought his input. In the writings below, he rarely mentions this part of his job. Yet it was what he spent almost half of his time doing; meeting quietly with individuals and congressional groups, with OMB officials, with national and international science leaders, and in this way he tried to influence the setting of government priorities.

Marburger, indeed, transformed the OSTP and its relation to the White House. Administratively, he streamlined the OSTP's management by appointing a single associate director for science and a single director for technology, instead of the statutory number of four such positions. Having four associate directors manage an office of only about fifty people, he thought, was top-heavy and expensive, encouraged operational stove-piping, and negatively impacted the number of scientists and engineers that the office could support. "I cannot imagine a situation where I would have wanted more than two Senate confirmed associate directors," he wrote at the end of his tenure.[1]

Operationally, Marburger both tightened and expanded the OSTP's focus. He tightened its focus by seeking to ensure that it focused on the science associated with policy issues without assuming that science was the only policy parameter. For example, climate change policy was heavily influenced by economic factors outside the core competency of OSTP. Likewise, he considered the embryonic stem cell debate to be primarily a question of ethics, not science policy. As he would tell his staff, there is no question about the scientific utility of embryonic stem cells. And while he had many

---

[1]John H. Marburger III, "Remarks on the Organization of Science Policy Advice in the Executive Branch," January 29, 2008, John H. Marburger III Collection, Box 38, Stony Brook University Special Collections and University Archives, Stony Brook University, NY.

conversations with Dr. Leon R. Kass, the chairman of the President's Council on Bioethics, Marburger did not have OSTP assume a prominent role in developing the guidelines for federal funding of embryonic stem cell research. Critics, however, saw Marburger as not pursuing energetically enough that part of the OSTP's mission to "ensure that the policies of the Executive Branch are informed by sound science." Nevertheless, changes in technology—which brought increased importance to such areas as control of the Internet—and Marburger's recognition of the need for OSTP to enter these areas resulted in an expansion of both the size and scope of the OSTP's White House role. Much of this expansion was because, while White House offices are usually barred from operational responsibilities, as an agency the OSTP was an exception. The OSTP, for example, had responsibility for national security and emergency preparedness communications, a responsibility that became significant after 9/11, and Marburger ran the Joint Telecommunications Resources Board, in charge of prioritizing the nonwartime emergency telecommunications resources. Marburger was also an active presence in several World Summit on the Information Society (WSIS) conferences sponsored by the United Nations, on international control of the Internet. At the time, Russia and China were attempting to wrest from the United States control over Internet standards, and specifically the Domain Name System (DNS). Ensuring international freedom of the Internet, Marburger felt, was the single biggest technology policy issue of the day, and his attendance and representation of the United States at key WSIS conferences played a major role in heading off the threat. The OSTP has maintained its expanded White House role ever since. This chapter focuses not on Marburger's role in specific policy issues but, rather, on how he addressed science and technology policy as OSTP director.

Having scientific and technological expertise in the White House was a powerful policy tool, Marburger realized, and the OSTP, with its formidable access to scientists and engineers, was an enormous reservoir of such expertise. But Marburger also realized that, for the OSTP to be maximally effective, he had to maintain the respect of the White House, cultivate the trust of other agencies and, in particular, the Office of Management and Budget (OMB), and prioritize projects based on advice rather than advocacy.

1. *Maintaining the Respect of the White House.* Working for George W. Bush, Marburger faced a huge challenge right from the beginning. One reason was

the popular image, actively promoted by the president's opponents, of George W. Bush as low-brow. Two days after the White House officially announced Marburger's nomination on June 25, 2001, he became a foil in comedian Jay Leno's opening monologue. Here's how the new science advisor was chosen, Leno said: Marburger had rubbed a balloon on Bush's head and stuck it to the wall—and the president was so impressed by this scientific feat that he decided on the spot to choose the man who could do that. The scientific community, too, regarded Bush with suspicion. Criticism began almost immediately, with some scientists saying that Marburger's cutting of the number of OSTP associate directors in half must have been dictated by the president and reflect a lack of concern for science. Another supposed indication of the Bush administration's lack of respect for science was that Marburger was not given the title of Assistant to the President, as had some previous presidential science advisors. This criticism was misplaced. In the Bush administration, that title was reserved for non-Senate-confirmed policy advisors; furthermore, Marburger much preferred the role of an advisor to that of an assistant. The criticism would mount over the next few years, focused on a small number of high-profile issues.

Marburger took seriously the notion that, as presidential science advisor and director of a White House office, he worked for the president. Just as Marburger had never said anything even vaguely critical about how Mario Cuomo worked, he also said nothing critical about George W. Bush, and he devoted executive time to cultivating a reputation as the president's impartial adviser. This frustrated many of his friends in the science community, who periodically approached him for the inside story of White House personalities and decisions. At one point Marburger drafted a brief note to himself outlining a response to such requests:

> I have often been asked to write or speak about aspects of my job as science advisor that have little to do with science and much to do with the specific work environment of my position. Most of these requests I have declined, and when I have accepted the invitation, I am sure I have disappointed my readers or listeners by speaking very generally, without personal details or anecdotes. The only time I speak concretely about workplace details is in response to questions from the audience following a talk, and sometimes in media interviews.

PRESIDENTIAL SCIENCE ADVISOR I   123

There are three reasons I remain silent on these details. First, they do not seem as important to me as the broader objectives of my work and the arguments for or against a specific course of action. Second, my specific work environment depends strongly on the specific people in that environment, and their personal styles and characteristics as they relate to me. Any statement I make about these characteristics is bound to affect the important personal relationships I depend on to get my work done.[2]

Marburger attended the 7:30 a.m. senior staff meeting in the Roosevelt Room of the White House, which was run by the chief of staff and included the rest of the president's senior advisors, such as the heads of the National Security Council, National Economic Council, and Domestic Policy Council and the director of the OMB. He was on good terms with the president, whose nickname for him—Bush was famously fond of nicknaming people—was the "Doc." Marburger also had access to cabinet members. He acquired influence in the White House over the areas that concerned him, as evident in his role in the American Competitiveness Initiative, discussed in chapter 5.

2. Trust of the OMB and other agencies.   Marburger also realized that his office had to cultivate the trust of the OMB. By itself, the OSTP does not control anything more than its own small agency budget of a few million dollars; it is the OMB staff members who write the actual numbers into the federal budget and therefore exert a huge influence on science policy. Many government agencies, he noted, tend to view the OMB as the enemy, for the actual budget figures generally end up less than requested. That view had the effect, in turn, of lowering the effectiveness of these agencies with the OMB. Marburger and his staff therefore spent time and care providing advice to queries from the OMB director and staff. By working closely with the agency and by providing honest advice, the OSTP became more integrated into the budget process and gained in influence as the trusted go-to place for virtually any science or technology question that arose in the OMB. "If the science adviser is disengaged from the Office of Management and Budget," Marburger told a New York Times reporter in 2005, "then he might as well

―――――――――――――

[2]Note, "JHM on Writing about Science Policy," June 26, 2006. John H. Marburger III Collection, Box 36, Stony Brook University Special Collections and University Archives, Stony Brook University, NY.

124   SCIENCE POLICY UP CLOSE

get on the lecture circuit."[3] Marburger's impact on this process is evident in
the title of the so-called Jack Daniels memo sent for several years to the
federal research agencies, written and cosigned by the directors of the OSTP
and the OMB. The memo outlines a set of science priorities for the budget
that the president will be putting together the following spring. The memo
is useful for the agencies when they submit their wish list of fundable projects, guiding them toward projects likely to end up in the president's budget, and therefore exerting influence on the character of those projects. It
was a highly clever way of getting the agencies, the OMB, and the OSTP on
the same page. Because "Jack" Marburger was OSTP director and Mitch
Daniels was OMB director, this annual budget memo was known as the Jack
Daniels letter, a nickname that stuck for a while even after Daniels's depar-
ture (he became governor of Indiana) from the OMB in June of 2003 and
Marburger's from the OSTP in January 2009. Marburger also contacted
other federal agencies besides the OMB, spoke with their leadership, and
asked them to create a mechanism in their agency to work with his office.
He traveled often to major U.S. research facilities and institutions to keep
abreast of what was coming down the pipelines. Marburger put into place a
structure that made it possible to provide up-to-date advice to the president, Congress, and federal agencies and successfully embedded the OSTP
more firmly in long-term technology policy.

3. *Prioritization on the Basis of Advocacy versus Advice.*   As a former university president and laboratory director, Marburger knew the brute necessity
to prioritize projects. Many people in the scientific community were uncom-
fortable with prioritization, feeling that all good science deserved to be funded.
But Marburger regarded his job as entailing the need to decide the priorities
within limited resources. In making such decisions, he felt, it was important
to consult the scientific community, but bearing in mind the difference between advocacy and advice. Advocacy he defined as "the practice of advancing arguments for certain actions without regard to the merits of competing
activities." Advocacy is legitimate and useful, he felt, for it provides policy
makers with important information. Advice he defined as "the practice of
identifying and objectively analyzing options for action and presenting the
result to decision makers," and he saw it as taking place in the context of a

---

[3]Daniel Smith, "Political Science," *New York Times,* September 4, 2005.

"larger framework of broad principles and objectives." Advocacy is "interest specific"; advice, "interest integrating." In public discourse, Marburger wrote, advocacy overwhelms advice and is hostile to prioritizing, but policy making depends on respecting the difference between advocacy and advice. "My most important responsibilities as science advisor to the president are to make that distinction, to identify principles that can be used to establish priorities across fields, and to manage a systematic process that discovers problems and opportunities, and develops action plans for the entire spectrum of technical endeavors in contemporary society."[4]

The evolution of Marburger's ideas about science and technology policy while presidential science advisor are best charted through a series of talks he gave each year to the American Association for the Advancement of Science (AAAS) at their annual Forum (initially called Colloquium) on Science and Technology Policy. The forum was established in 1976, the same year as the OSTP itself, and meets in Washington, DC in the spring. It brings those who practice science policy at the highest level together with those who are affected by it, and it analyzes the president's budget request related to science and technology. By tradition, the presidential science advisor delivers the keynote address. Marburger wrote each of these speeches himself, and he valued the occasions more than many government officials who engage in public speaking. He did not treat the speeches simply as an opportunity to announce or propound administration policy. He regarded them, first, as a valuable opportunity to present the key science policy issues facing the administration, and second, as a valuable exercise on how to put forward the best argument on these issues to an educated, analytical, and critical audience. While in other speeches Marburger might discuss specific scientific or technological issues that he was confronting—such as the International Space Station, the International Thermonuclear Experimental Reactor, stem cells, climate change research, and control of the Internet—in the AAAS policy forum addresses he focused on broader science policy. This chapter reprints the policy forum talks that he gave during the first term of the Bush administration, while the next chapter reprints those delivered during the second term. Each talk is interesting in itself for different reasons;

---

[4]"Four Themes in Science Policy," paper presented at the Brookings Institute Policy Conference, Washington, DC, June 11, 2007. John H. Marburger III Collection, Box 37, Stony Brook University Special Collections and University Archives, Stony Brook University, NY.

together they provide a comprehensive picture of Marburger's approach to science policy during his years with the Bush administration.

—RPC

American Association for the Advancement of Science
27th Colloquium [Forum] on Science and Technology Policy
Keynote Address no. 1
Washington, DC
April 11, 2002

Marburger's first AAAS policy forum keynote address, exactly seven months after 9/11, was dominated by science policy in connection with the response to terrorist attacks and by the changing role of the OSTP in the wake of these events thanks to heightened attention to communications technologies and bioterror. Key issues broached concern the role of antiterrorism vis-à-vis other forces driving science policy, the distinction between "science-based" and "issues-based" policy, balance and prioritization in federal science funding, the importance of the science infrastructure, the distinction between "discovery-oriented" and "issue-oriented" science, the need to cultivate the OMB, the importance of the social sciences to science policy, and the importance of the science and technology workforce.

—RPC

Thank you very much for inviting me to speak today about the administration's science and technology strategy. I received the invitation just hours before the Senate confirmed my appointment last October, and at that time I had only the dimmest notion of how our strategy would be shaped by the terrorist attacks of the month before. My first actions were to structure OSTP to serve the president in the war against terrorism and to reach out to the science and higher education communities with a call to action. Two seasons have now passed, the war has progressed in stages both abroad and on the homeland front, and we have had time to compare the needs of antiterrorism with the other forces driving science policy. This colloquium comes at an ideal time to take stock and reflect

PRESIDENTIAL SCIENCE ADVISOR I 127

on what has happened, and what a reasonable future course might be for the nation's science policy in this vulnerable world.

OSTP Today

In the Bush administration the Office of Science and Technology Policy continues to play a strong role in shaping science policy. The interagency coordinating mechanism of the National Science and Technology Council has proved important, not only for the integration of agency actions in the war against terrorism but also as a nucleus for the crystallization of agency expertise needed for immediate response to urgent issues. Early service to the Office of Homeland Security following the exploitation of the U.S. mail for bioterrorism last fall was the first in a series of new activities somewhat different from the historical OSTP norm. Today OSTP manages the research and development needs of Homeland Security through a shared senior staff member who has access to all the OSTP technical resources in a form of matrix management. The entire office is now organized to provide focused advice when and in the form needed by the agencies and policy offices we support. The NSTC [National Science and Technology Council] interagency task forces and working groups continue to function as before, with less emphasis on analysis by OSTP staff and more on seeking strategic input from other appropriate organizations including the agencies and the National Academies. The President's Council of Advisors on Science and Technology (PCAST) has been formed and is functioning through a series of panels, one of which focuses on terrorism, others on energy, telecommunications, and science funding policy. I have endeavored vigorously to reconnect the office with the Academies, the science and engineering societies, higher education, and the technically oriented private sector. We are working closely with the Office of Management and Budget to implement the president's management agenda and to shape funding policies to meet the rapidly changing conditions of contemporary science. Today OSTP is active, engaged, and effective in the formation and execution of the nation's science policy.

Initial Responses to the War against Terrorism

My message to the science and higher education communities last fall was, first and foremost, to appreciate how deeply committed the president is to winning the war against terrorism. That commitment includes the mobilization of every sector, including science, engineering, and higher education. Shortly after the September 11 attacks, many federal agencies launched initiatives to respond to terrorism issues and funded them with existing appropriations. The Department of Defense and the State Department created a Combating Terrorism Technical Support Working Group that solicited, evaluated, and funded specific projects that could improve technology needed for this special kind of war. Some needs were obvious, such as increasing the capability of first responders to detect bio- or chemical hazards, better ways to sift intelligence data from multiple sources, and better vaccines and therapies for biopathogens. Other needs are more strategic: defining and assessing the nature of terrorist threats, or analyzing and strengthening the nation's logistical infrastructures in transportation, communication, energy distribution, food supply, and health care. Many of us realized that these longer-term issues would require considerable thought and consultation with the nation's intellectual community. To this end, the National Academies sponsored an important meeting late in September to consider how they might organize science input to the war effort. I learned much from that event and agreed to establish an interagency task force that would take up recommendations produced by an NAS committee proposed at the meeting. The committee, cochaired by Lewis Branscomb and Richard Klausner, is likely to produce useful guidance by midsummer.

The institutions that produce science and technology are not only sources of solutions and advice; they are also potential targets and means of exploitation for terrorism. Universities can inadvertently provide materials, skills, and concealment for terrorist operations. They cannot ignore their responsibility to society for limiting the opportunities for such perversions of their educational and research missions. Universities need to think through these responsibilities and advise governments where to draw the line between avoiding terrorist risk and obstructing the processes of education and discovery. During the weeks following September 11, I met with higher education leadership organizations to urge them to begin dialogues on their campuses to define their positions on terrorism and to clarify where the balance must be struck in response to society's desire to protect itself. OSTP is fostering and closely monitoring the broader dialogue on these issues within the administration.

Innovation versus Implementation in the War against Terrorism

As I learned more about the challenges of terrorism, I realized that the means for reducing the risk and consequences of terrorist incidents were for the most part already inherent in the scientific knowledge and technical capabilities available today. Only in a few areas would additional basic research be necessary, particularly in connection with bioterrorism. By far, the greater challenge would be to define the specific tasks we wanted technology to perform and to deploy technology effectively throughout the diffuse and pervasive systems it is designed to protect. The deep and serious problem of homeland security is not one of science; it is one of implementation.

This has two consequences. First, terrorism is not going to be a significant driver for science funding in general. Second, those seeking federal agency customers for specific technology products are likely to be frustrated while implementation strategy is being developed. Unlike conventional warfare, where trained personnel employ purpose-built technology in a localized and well-defined threat environment, the war against terrorism is waged across all society against the vulnerabilities of poorly defined and fragmented systems, only a few of which are owned or controlled by the federal government. Central authorities have limited means available to reduce risky practices in the private sector, and often the simplest means are unacceptably intrusive to American society.

Science-Based Science Policy

I do not mean to imply that the science role in the war against terrorism is unimportant, nor that substantial funds will not be forthcoming to solve technical problems critical to the war effort. But science is moving forward with its own powerful dynamic, and it is producing the means for addressing many of society's difficult problems, terrorism among them. Federal support of science must be directed first of all to sustain this dynamic, and second, to seize the greatest opportunities it is creating for discovery and for the improvement of the human condition. This is what I have called a "science-based" science policy. It differs from what might be called an "issues-based" policy in recognizing that discovery and the creation of entirely new technologies are unlikely to emerge from mandates in service to a particular social issue.

When we say that necessity is the mother of invention, we ought to admit that this normally applies to a sort of Edisonian invention that exploits existing science and ripe technology. The scope of existing science is defined by the limits of existing technology. New science requires advances in technology that are not obviously relevant to any social need. And yet each such advance has contributed substantially to social improvements. Many examples come to mind: lasers, sensing and imaging technologies, and superconductivity. This is a familiar argument for societal support of basic science. But it is not the main point of the science based policy that I am advocating. The main point is that science has its intrinsic needs and processes that have to be supported if the whole apparatus is to work effectively. If we ignore these needs and direct funding according to the severity of social problems we would like science to address, we tend to enrich only one part of the machinery and diminish our ability to address the critical problems.

Balance in Science Funding

This is my way of discussing the problem of "balance," sometimes expressed as too little funding for NSF [National Science Foundation] compared with NIH [National Institutes of Health], or as too little for

PRESIDENTIAL SCIENCE ADVISOR I   131

the physical sciences compared with the life sciences. I think "balance" is a misleading term for the real issue, and it is a dangerous term. Elsewhere I speculated, to make this point, that perhaps the recent large increases for NIH have simply enabled health researchers to exploit the same fraction of opportunities for discovery in their field as physical scientists can in theirs under existing budgets. A strong case can be made that the discovery of the molecular basis of life processes created research opportunities vastly greater than those in the physical sciences. It is not so much the balance among fields of science that is problematical; it is the balance among the different parts of the machinery of science.

We are witnessing advances in the technical infrastructure of science that do justify large increases in certain fields, and large increases have been forthcoming. The president's proposal to complete the doubling of the NIH budget in fiscal year 2003 is an example. So are the priorities given to information technology and nanotechnology. I am not the first to declare the revolutionary nature of these advances, which are occurring in instrumentation that permits us to image and control the properties of organic and inorganic materials at the atomic scale, and in powerful computational and information technology that enables us to manipulate atomic level information and to simulate the functional properties of matter based on its atomic-level description. The result is an unprecedented understanding of, and power to control, the functional properties of all matter composed of chemical atoms. One product of these capabilities is the current excitement of biotechnology and its inorganic counterpart, nanotechnology. Both deserve support in proportion to the potential they hold for discovery. My impression is that potential is greater in the vastly more complex organic domain, but huge opportunities remain to be exploited in both areas.

This oversimplified revolutionary scenario has four central components: two enabling technologies of instrumentation for atomic-scale imaging and control, and computation for simulation and management of the atomic-scale information, and two discovery-oriented fields of biotechnology and nanotechnology that have been empowered by these tools. Only one component, instrumentation, is not currently identified as a priority in the federal budget. Attention to instrumentation will require analysis of the related issues of aging or inadequate facilities. Because some of the instrumentation dwells in the domain of "big science" (for example, synchrotron x-ray sources, or nuclear reactors), this issue is also related to the funding of science within the national laboratories. If we want to achieve balance in federal science funding, we are going to have to understand how the complicated funding process works, or fails to work, to sustain the essential tools upon which our most exciting and productive areas of science and technology depend. Once the quality of this infrastructure is assured, then the questions of priority and adequacy of funding for the dependent fields remain. These questions must be answered in different ways for the two types of research that are typically described as basic and applied. I call them discovery oriented and issue oriented. Priorities for issue-oriented science are driven by the nation's policy agenda. For discovery science, priorities must be consistent with the opportunity for discovery, and that is a matter for the experts.

The President's Management Agenda Applied to Science

Scientists do, of course, make judgments all the time about promising lines of research. Scientists choose lines to pursue, and they choose lines to drop as unproductive. That is the only way they can remain productive and relevant in their scientific careers. Some make those choices better than others. It makes sense for the world's largest sponsor of research, the U.S. government, to want to make such choices as wisely as the most productive scientists do. This, in my opinion, is how to think about the president's management agenda as it applies to science. I shared the podium last month with budget director Mitch Daniels at a National Academy workshop on this subject and witnessed much nervousness in the audience about the prospect of evaluating basic science. But is it possible to decide rationally when to enhance or to terminate a project if we do not possess a way of measuring its success? Most program officers within the science-funding agencies insist on peer reviews and peer judgments of the projects they are funding, and undoubtedly peer review will remain an essential part of evaluation. I think peer reviewers apply criteria in coming to their judgments, and I think the process of evaluation would be more credible if those criteria could be made explicit. That is why I support OMB's effort to introduce more specificity in the science evaluation process. Good advice about the principles that should guide evaluation has been produced by the National Science Board and the National Research Council. More needs to be done to adapt this process to the different parts and different ways of doing science.

The Importance of the Social Sciences

Management and evaluation are activities that can be studied objectively and improved systematically with the tools of social science. As a university president, I was always puzzled by how rarely academic managers took advantage of their own disciplines in dealing with their departments. Industry seems to have a more academic approach to management than higher education does. The social sciences in general have much more to offer on the difficult problems of our time than we are currently acknowledging in our federally funded programs. The September meeting on terrorism sponsored by the National Academies included a number of social scientists whose input provided structure and dimension to the discussion. We are not yet systematically including the social sciences in the mobilization for the war against terrorism, and this needs to be done.

I do not completely understand why we have failed in the past to develop and use the social sciences more effectively as a tool for public policy. Perhaps here, too, we have not paid enough attention to the structure of the field itself and what it needs to function well. Social science also possesses the three tiers of infrastructure, discovery science, and issue-driven science, and agency programs need to reflect these more explicitly. There is no doubt that the social sciences suffer from treating issues that are so familiar as to breed contempt.

Workforce Issues

No issue deserves more attention from the social sciences than that of the future of the technology workforce. Many observers have expressed concern that economic globalization is creating attractive opportunities for employment in the nations on whom we have been relying for imported technical talent. What would happen if all the foreign graduate students returned to their countries of origin immediately after receiving their degrees? A catastrophic loss of technical capability would ensue. Already many industries are having difficulty recruiting technically trained personnel.

The prospect of such a catastrophe is one motivation for a widespread interest in education, especially math and science education. President Bush cares passionately that children acquire skills systematically throughout their education that will prepare them to participate in the twenty-first century workforce. Many factors necessary to this objective are receiving attention, from achievement testing to teacher recruitment and training. Educational research programs are being overhauled or strengthened, and initiatives funded to enrich science and math teaching through partnerships between schools and universities.

I believe the workforce problem is more complicated than we have yet acknowledged and will be difficult to define, measure, and resolve. The market for intellectual talent has been a global one for many years and will continue to be in the future. How we ensure national security and national economic competitiveness in such an environment is an unanswered question. Surely we will need wise advice on this issue from the social science community.

I wish to thank the association for making it possible for me to reflect on these issues at such a high level of abstraction. Others in this colloquium will provide considerable detail on nearly all the issues I have discussed this morning. I look forward to hearing and reading their contributions.

American Association for the Advancement of Science
28th Colloquium [Forum] on Science and Technology Policy
Keynote Address no. 2
Washington, DC
April 10, 2003

Marburger realized that the scope of science policy went far beyond settling budgets; it also involved pressing the case for science funding and establishing what fraction of the science budget each science agency would get. Science policy must also include ensuring the efficiency of the scientific infrastructure so that policies get implemented successfully. "Science has its intrinsic needs and processes," he remarked in his first AAAS policy forum address, "that have to be supported if the whole apparatus is to work effectively." Ignoring these needs would "diminish our ability to address the critical problems." In his second AAAS policy forum address, Marburger focused on a single infrastructure issue: the perceived crisis in U.S. science and technology manpower brought about by the increased restrictions that Homeland Security had placed on foreign scientists and students after 9/11, resulting in a backlog of visa applications, among other problems. This issue—the negative impacts of security measures on science—was not only an issue that was affecting science research and development on a truly global level, but was also rapidly moving from an annoyance to a lightning rod for the scientific community's ire toward the administration, threatening to become a political issue. Marburger's address is a model for how to handle such a controversy. Seeking to defuse conflict by informing, Marburger carefully researched the issue and walks his audience through his findings: the regulations, the process, the statistics—including changes in the number of visa applications—the number of approvals, rejections, and cases submitted for additional review, and the changes in the number of institutions certified to admit foreign students. He frequently instructs his audience where to find other, more detailed information about these topics if they have further questions. This comprehensive, fact-based review enables him to say, "We think we understand what is happening, where the problems are, and how they can be addressed." Marburger concludes by expressing frustration with the dominance of anecdotes over information, and by outlining with a series of "general principles" that will help to resolve the problem, whose implementation will require not only changes in the administration's practices but also those of the affected communities, suggesting that good federal science policy requires not only governmental leadership but also the conscientious participation of those affected.

—RPC

This annual colloquium provides an opportunity for me to speak broadly about the policies motivating science programs and budgets within the administration. My office has been very active during the past year working with PCAST [President's Council of Advisors on Science and Technology], interagency committees, and other groups to identify issues and form policy responses to them. During this period my OSTP colleagues and I have testified frequently before Congress and addressed many conferences and symposia such as this one. This administration values science and technology and believes it underlies its priorities of homeland and national security and economic vitality. We have attempted to be clear about priorities and to make them widely known.

In view of the widespread availability of information about science policy and priorities, I decided to narrow my remarks this morning to a single important issue affecting the science and higher education communities. The issue is the ability of foreign technical personnel, including students and scientists, to visit the United States for meetings, research collaborations, or educational pursuits. This week's *Chronicle of Higher Education* has a special section titled "Closing the Gates: Homeland Security and Academia" with articles focusing on the issue. The thrust of the articles is that in our determination to protect the homeland, America is cutting itself off from the vast benefits foreign students and technical personnel bring to our country.

Let me begin by stating clearly that this administration values the contribution foreign scientists and students make to the nation's scientific enterprise, to our economy, and to the appreciation of American values throughout the world. We want to make it possible for any visitor who does not mean us harm to come and go across U.S. borders without significant inconvenience or delay. We believe it is possible to take appropriate precautions against terrorism without inhibiting the numerous relationships with other nations that are essential in today's globalized technical society. And we mean to apply ourselves to the development of efficient ways of taking these precautions until that goal is achieved.

My purpose this morning is to review the "visa situation" and attempt to clarify current policy and current actions that are being taken to achieve that policy. I will not talk at all about monitoring of foreign visitors once they are in the country, and I will say little about the details of the Student and Exchange Visitor Information System (SEVIS). My aim is to characterize the visa system as it applies to students and scientists and describe what is being done to make it work better. Just two weeks ago, on March 26, the State Department's deputy assistant secretary for visa services, Janice Jacobs, testified on these issues before the House Science Committee. I commend her testimony to you as a current authoritative source of information.

The visa is a travel document that permits someone to reach the U.S. border and seek admission. The Department of State administers the visa process. Admission to the country is determined by the immigration border inspectors of the new Department of Homeland Security, to which this responsibility was transferred by the Homeland Security Act of 2002. In general, the visa process remains essentially the same today as it was prior to 9/11. One important new provision is the statement in Section 306 of the Enhanced Border Security and Visa Entry Reform Act of 2002 (Public Law 107–173) that "no nonimmigrant visa . . . shall be issued to any alien from a country that is a state sponsor of international terrorism unless the Secretary of State determines, in consultation with the Attorney General and the heads of other appropriate U.S. agencies, that such an alien does not pose a threat to the safety or national security of the United States." The State Department has also made some changes to the process under the existing authorities of the Immigration and Nationality Act, section 212 (1182 in the U.S. Code), which I will describe later. Also, shortly after 9/11 (October 29), President Bush issued Homeland Security Presidential Directive 2 (HSPD2), which among other things called for enhanced immigration enforcement capability and an end to abuse of international student status. To quote from that directive:

The United States benefits greatly from international students who study in our country. The United States Government shall continue to foster and support international students. The Government shall implement measures to end the abuse of student visas and prohibit certain international students from receiving education and training in sensitive areas, including areas of study with direct application to the development and use of weapons of mass destruction. The Government shall also prohibit the education and training of foreign nationals who would use such training to harm the United States or its Allies.

The directive calls for the creation of a program that "shall identify sensitive courses of study, and shall include measures whereby the Department of State, the Department of Justice, and United States academic institutions, working together, can identify problematic applicants for student visas and deny their applications." This is the directive for which the process known as IPASS was devised and presented to the higher education community last year. I described IPASS in testimony to the House Science Committee on October 10, 2002. This Interagency Panel on Advanced Science and Security would provide systematic input from scientific experts to define and identify the "sensitive areas" mentioned in the presidential directive. The legislation creating the Department of Homeland Security changed the status of the agencies participating in the original IPASS process, and new roles are not yet completely defined. Consequently, IPASS is not yet in place.

There is a potential downside to IPASS, which you will understand as I move on to describe the current visa situation. The worst aspect of the situation is the long delays in processing some visa applications. If IPASS adds even more steps to the process without adding value, it may increase wait times, which is not our intention.

First, let me give you some numbers. For the past four years, the annual number of nonimmigrant visa applications has varied between 8 and 10 million, of which about 75 percent are granted. There are multiple attempts per individual, so the actual success rate of individuals is higher. Of those admitted, approximately 20 percent are in the F, M,

PRESIDENTIAL SCIENCE ADVISOR I 139

and J categories in which students and exchange visitors fall. In 2000, for example, those admitted in these categories totaled about one million individuals. Prior to 9/11, an estimated 75,000 institutions were certified to admit foreign students. This number has dropped to about 8,000 today. The large decline is attributed to English language and small vocational schools that are no longer in business. DHS [Department of Homeland Security] has adjudicated all timely and complete applications for recertification. By this August all international students must be registered through SEVIS. National laboratories and other institutions also use SEVIS to enter and track foreign visitor information.

Congress had mandated an automated foreign student tracking system in section 641 of the Illegal Immigration Reform and Immigrant Responsibility Act of 1996, responding to the first World Trade Center bombing. A pilot project began in the INS [Immigration and Naturalization Service] Atlanta district the following year, but further implementation was slow. After 9/11, the pilot project was converted to SEVIS. The USA Patriot Act of January 2003 required SEVIS to be implemented by January 1, 2003. The system has experienced well-publicized glitches. DHS has hired experts to identify and resolve issues and is monitoring and correcting problems, a process that will continue as long as necessary. We have to make this system work, because it provides information to decision makers at multiple steps in the visa control process.

The visa process begins at the consular office in the country of origin, and the first challenge for applicants is filling out the forms correctly and submitting them well in advance of the intended visit. It appears that expectations for the accuracy and completeness of the visa application forms and the accompanying I-20 pertaining to students have increased since 9/11, and I think most would agree that is appropriate. Consular officers judge each application on its merits in accordance with existing immigration laws and procedures. The Immigration and Nationality Act (INA) lists specific grounds for ineligibility, each of which must be considered by the officers. The first opportunity for rejection comes at this stage, and the most common cause by far is "failure to establish intent to return to the home country." This refers to Section 214(b) of the INA. The law presumes that a nonimmigrant applicant intends to immigrate until he/she can demonstrate otherwise. The burden of proof is on the applicant to show compelling ties outside the United States that will cause the applicant to leave the United States after the authorized temporary stay. Examples of what kind of proof is necessary are provided as guidance to applicants, and they do not appear to be unreasonable. I have heard complaints that this particular provision of the INA is contrary to our desire to capture the most talented students into our domestic workforce. But there are, and clearly should be, different visa categories for those who intend to immigrate versus those who merely intend to study. Student visas are not immigrant visas or temporary worker visas, and applicants should be aware of this.

The next largest cause of rejection at the consular stage seems to be "Application does not comply with INA requirements." As far as I can tell from the data I have seen, no other category of rejection comes within orders of magnitude of these top two. Generally, the latter category covers denials pending receipt of additional documents or interagency security clearances. The INA contains several additional grounds of ineligibility, including provisions based on national security grounds. I have personally reviewed detailed statistics for rates of acceptance and rejection over the past five years in various visa categories from various countries and found a small but significant decrease in acceptance rates over all categories. Changes in student and scientist rates do not appear to differ from those of other categories.

So where is the problem? Unfortunately, while rejection rates for science- or study-related activities remain small, the number of cases submitted for additional review has increased dramatically since 9/11. This increase, plus more careful scrutiny of the submitted cases, has led to processing backlogs that have created excessive delays in notification. Solutions focus on removing these backlogs and changing the way cases are processed, without sacrificing the rigor of the review.

Three different review procedures dominate the process for the classes of visas we are considering. First, all applications are checked with the Consular Lookout Automated Support System (CLASS). This system compares names with lists from the FBI's National Criminal

Information Center, the intelligence community's TIPOFF database on terrorists, and so on. If a "hit" occurs, then the consular officer must take certain steps. In some cases, the application must be sent to Washington for further review. These reviews seem to be resolved within thirty days in nine cases out of ten.

The other two reviews are conducted only when the consular official judges that the application meets special criteria. One of these, code-named MANTIS, was established under section 212 of the Immigration and Nationality Act, which I mentioned earlier. The purpose of this section is to exclude applicants whom a consular official or, since March 1, the Secretary of Homeland Security has reasonable grounds to believe intends to violate or evade laws governing the export of goods, technology, or sensitive information. The decision to submit an application for MANTIS review is based on guidance accompanying a Technology Alert List compiled by State Department Officials with input from other federal agencies. The other federal agencies also assist in evaluating the cases. In August 2002, the guidance for the TAL was clarified for each category. The statistics tell a story: In calendar year 2000, about 1,000 cases were reviewed under MANTIS, and 2,500 the following year. In 2002 the figure jumped to 14,000, overloading the system last summer and fall. Today the State Department estimates that at any given time there are about 1,000 visa applications in the MANTIS review process. FBI and State are dedicating full-time individuals to clean up the backlog. Janice Jacobs's testimony pointed to twelve new personnel, cross-training of existing personnel, and management and technology improvements as evidence of State Department efforts to reduce the backlog.

The second special review, code-named CONDOR, is entirely new since 9/11 and is devoted to identifying potential terrorists. In both systems, the flood of new case submittals following 9/11 required changes in methodology to maintain the quality of the reviews. In the past, if the State Department received no derogatory information from the supporting agencies in thirty days, it was assumed there was no objection to the visa issuance. But in the summer of 2002, the backlog was such that no agency could give assurance that thirty days was enough, and the thirty-day rule was suspended. State must now wait for affirmative replies from participating agencies before it informs consular officers that there is no objection to issuance.

My aim in going through this process is not to make you more discouraged but to give you hope that the visa situation can be improved. We think we understand what is happening, where the problems are, and how they can be addressed. My office, working closely with the Homeland Security Council, has had good cooperation from the Departments of State, Justice, and Homeland Security, all of whom agree that improvements are needed. And there have been notable successes, including cooperation last fall among six offices and agencies to identify and resolve inefficiencies and duplications in the CONDOR process that cleared out nearly 10,000 applications from the backlog. The same group is now working on similar issues in MANTIS. Part of the problem is that all these systems of special review operate in parallel but impact the same offices. So backlogs in one system can affect flow rates in all the others. That is why applications for short-term visits, including B visas, are held up along with all the others.

I have organized my office to place special emphasis on homeland security issues, including visa issues. Between October 2001 and March 2003 OSTP played an unusual operational role in supporting the Office of Homeland Security. The existence of the new Department of Homeland Security makes it possible for us to focus on our traditional role of coordination, oversight, and policy formation. Within OSTP, homeland and national security functions now report through a senior director, Bill Jeffrey, directly to my chief of staff and general counsel, Shana Dale. The visa situation is one of a small number of issues that has received top priority for the past eighteen months. We see solutions emerging, but they involve multiple agencies and large volume processing systems. The stresses resulting from the events of 9/11 cannot be relieved as rapidly as they emerged. But I am optimistic that they will be relieved.

Some general principles that will help this process include the following.

First, increased and systematic involvement of the expert communities within the federal government in providing guidance to the process. Whether it is crafting the Technology Alert List, or helping consular officials to employ it, or reviewing cases in the CONDOR and MANTIS systems, or IPASS, technical expertise is essential. IPASS, not yet implemented, could be a model for embedding technical expertise.

Second, elimination of duplicate operations among the three screening processes, CLASS, CONDOR, and MANTIS. The Departments of Homeland Security, State, and Justice have already made very beneficial adjustments in methodology, and I expect to see more. Gains here have a nonlinear impact on throughput because of the interaction among the different systems.

Third, continued improvement of impact reporting from affected institutions. We have many anecdotes, but they do not add up to trends, and they give little insight into where or how the systems can be adjusted to best advantage.

Fourth, better knowledge among all parties regarding how the visa system works and what are its objectives. Very few applicants are terrorists, and therefore a properly working system will not reject large numbers on grounds related to terrorism. It should, however, be rejecting some, and that is happening. Most of the current delays and backlogs are related to our efforts to screen applicants more rigorously, and not as the result of policies to exclude. Knowing more, we can advise visa applicants better. Students and visiting scientists need to get accurate information from their institutions and collaborators about how and when to apply for visas. We can all help make the system work better.

Fifth, a frame of mind within the technical and higher education communities that perhaps falls short of patience but rises above hysteria. We are facing a serious challenge, and this administration is responding seriously to it. We have evidence of cooperation among agencies and appreciation for the importance of the task. If the devil is in the details, then so is the opportunity for good will to produce a favorable outcome.

144   SCIENCE POLICY UP CLOSE

American Association for the Advancement of Science
29th Forum on Science and Technology Policy
Keynote Address no. 3
Washington, DC
April 22, 2004

Marburger's third AAAS forum keynote address took place during the final
year of George W. Bush's first term in office—thus potentially Marburger's
final opportunity to address the forum as presidential science advisor. In it,
he summarizes the science funding profile of the Bush administration over
the previous years and explains the various factors that had shaped this
profile: the rapid growth of technology, the emergence of worldwide terrorism, bioterrorism, nanotechnology, and the visa problem. One of the most
noteworthy features of this talk, however, is the way Marburger handles critics in the scientific community who had accused the Bush administration of
undermining the integrity of science through politically motivated actions.
This requires some background. On February 18, 2004, in an effort organized by the Union of Concerned Scientists (UCS) at the beginning of an
election year, a group of sixty-two prominent U.S. scientists released a
statement titled "Restoring Scientific Integrity in Policy Making." The statement, whose signatories included twenty Nobel laureates, charged the Bush
administration with "suppressing, distorting or manipulating the work
done by scientists at federal agencies," in areas including pollution, climate
change, endangered species, and evidence that Iraq was seeking uranium
enrichment centrifuges for an atomic bomb project.[5] The signatories included several personal friends of Marburger's, including David Baltimore,
a Nobel laureate in medicine who was president of the California Institute of
Technology; Leon Lederman, a Nobel laureate in physics who was the former president of the AAAS and former director of Fermilab; and Sidney
Drell, a physics professor at the Stanford Linear Accelerator Facility. The

---

[5]"Scientific Integrity in Policymaking: An Investigation into the Bush Administration's Misuse of Science" (Union of Concerned Scientists, 2004), 5, www.ucsusa.org
/assets/documents/scientific_integrity/rsi_final_fullreport_1.pdf. The accompanying
statement, titled "2004 Scientist Statement on Restoring Scientific Integrity in Policy
Making" (www.ucsusa.org/scientific_integrity/abuses_of_science/scientists-sign-on
-statement.html), was later signed by more than 1,500 scientists.

PRESIDENTIAL SCIENCE ADVISOR I

statement was a political document that crystallized and fomented anger within the scientific community against the administration. Most notoriously, Howard Gardner, a cognitive psychologist at Harvard, said in an interview on National Public Radio, "I actually feel very sorry for Marburger, because I think he probably is enough of a scientist to realize that he basically has become a prostitute." In a later interview with the New York Times, Gardner expressed some regret about the remark, saying, "I wish I'd used it as a verb rather than as a noun." Marburger refused to reply, displaying an objectivity toward a scenario that included such personal insults. As the White House steward for U.S. science policy, Marburger set out to craft a reply to the "Restoring Scientific Integrity" statement, taking its charges, as he says in this address, "very seriously." Indeed, as he continues, "as director of the Office of Science and Technology Policy, I accept the responsibility to hear and respond to issues affecting the integrity of technical advice within this administration." Marburger's response was twofold: first, he treated the statement not as a political document or as political theater but as a series of factual claims; second, he stressed the positive aspects of science policy under the Bush administration and the negative impact on science of the approach taken by signatories of the letter. Marburger and his staff itemized the document's charges and investigated in each case whether there was an issue to be dealt with. He contacted several signers of the document, and discovered that some had not read the document—and that still more had not investigated its claims—before agreeing to affix their name to it. In a report issued April 2, he concluded that many—though not all—of the incidents were distorted and that in any case they did not indicate a systematic antiscientific behavior on the part of the Bush administration. Most important, he thought that the more partisan the scientific community got, the less effective it became. "I am not saying 'everything is fine,'" Marburger wrote Sidney Drell. "I am saying that identifying a common cause of all these things is wrong. It is not only wrong, but counterproductive to any effort to make things work better."[6] The UCS countered Marburger's response on April 19, reiterating many of the charges. The UCS and other critics were making an effort to depict the Bush administration as antiscience in the

---

[6]John H. Marburger to Sidney Drell, e-mail of April 9, 2004, John H. Marburger III Collection, Box 38, UCS folder, Stony Brook University Special Collections and University Archives, Stony Brook University, NY.

months preceding the election. In this address, Marburger does not dwell on this controversy, preferring to point to objective measures of how well science has fared in the first term of George W. Bush's administration. "You really have to work at it to make a counterargument that science has not fared well in this administration," Marburger told a reporter.[7] The upbeat character of this address, only three days after the latest salvo, and its focus on facts and comparisons with the past show off Marburger the policy actor at his best in a difficult situation. So does his primary focus here on his aim "to assess current status and look ahead to future policy in view of past experience."

—*RPC*

This is my third appearance as presidential science advisor in this annual Forum on Science and Technology Policy, and the last in the current presidential term. When this forum took place in 2001, I was (in retrospect) blissfully engaged with a diminishing menu of management and environmental issues at Brookhaven National Laboratory and enjoying an outpouring of excellent science from Brookhaven's facilities and their users. I came to Washington on September 15, 2001, and participated in the AAAS symposium on terrorism in December. At the 2002 policy forum, I presented my thoughts on future priorities in science and technology funding and related them to intrinsic opportunities in science and to current deep movements in society. In 2003, I addressed the single most serious issue facing science in the aftermath of the terrorist attacks of September 11, 2001: the backlog of visas for students and scientists, and related problems. Now, in 2004, the main external factors driving policy in this administration are well established—particularly in the areas of national and homeland security and the economy—and it is a good time to assess current status and look ahead to future policy in view of past experience.

---

[7]Jeffrey Mervis, "Bush Victory Leaves Scars—and Concerns about Funding," *Science* 306 (November 12, 2004), 1110–1113.

Administration Priorities and R&D Budgets

President Bush has made it abundantly clear that his budget priorities have been to protect the nation, secure the homeland, and revitalize the economy. His budget proposals to Congress are in line with vigorous actions in each category. Increases in expenditures for homeland security, in particular, have dominated changes in the discretionary budget during this administration, and we have seen the emergence of a significant new science and technology agency within the Department of Homeland Security (DHS). The current budget proposal for the DHS science and technology function is $1.2 billion, with an estimated total of $3.6 billion in homeland-security-related R&D in all agencies. The science and engineering communities exerted a significant influence on the structure of the new department, particularly through the National Research Council report *Making the Nation Safer.*

Each of the three overarching presidential priorities has strong science and technology components. The president has sought, and Congress has appropriated, substantial increases in research and development budgets not only for homeland security but also for defense and for key areas of science and technology related to long-term economic strength.

R&D expenditures. R&D expenditures in this administration are up 44 percent over the past four years to a record $132 billion proposed for 2005 compared to $91 billion in fiscal year 2001, and the nondefense share is up 26 percent. The president's fiscal year 2005 federal R&D budget request is the greatest share of GDP in over ten years, and its share of the domestic discretionary budget, at 13.5 percent, is the highest level in thirty-seven years. Nondefense R&D funding is the highest percentage of GDP since 1982. Total U.S. R&D expenditures, including the private sector, were at 2.65 percent of GDP in 2002, the most recent year for which I have data. I suspect it is above that today. Its historical high was 2.87 percent in 1964 as NASA was ramping up for the Apollo program.

Nondefense R&D spending. The fiscal year 2005 request commits 5.7 percent of total discretionary outlays to nondefense R&D, the third highest level in the past twenty-five years. While the president has proposed to reduce the overall growth in nondefense, non-homeland security spending to 0.5 percent this year to address overall budget pressures, his budget expresses a commitment to "nonsecurity" science with a considerably higher growth rate at 2.5 percent.

Basic research. During the current administration, funding for basic research has increased 26 percent to an all-time high of $26.8 billion in the fiscal 2005 budget request. What Congress will do with the presidential requests for science is at this point an open question. I do want to acknowledge that Congress has treated science well in its appropriations, and the good figures for science during this administration represent a strong consensus between the legislative and executive branches that science is important to our nation's future.

As I emphasized in 2002, priorities for these large expenditures respond to two important phenomena that have shaped the course of society and are affecting the relationship of society to science, namely, the rapid growth of technology, particularly information technology, as the basis for a global economy, and the emergence of terrorism as a destabilizing movement of global consequence.

Science and Security

I have been speaking of these phenomena ever since I arrived in Washington more than thirty months ago. The good news is that the accelerating pace of technology and its obvious economic impacts have captured the attention of governments at every level, creating an awareness that science is the source of technology and a consensus that economic vitality is a strong rationale for public support of science. The bad news is that the new technology-intensive infrastructure of society makes it vulnerable to terrorism, and sensible responses to terrorism can have negative consequences for the conduct of science.

Balancing science and security has become a major theme of science policy during my tenure in Washington. I believe it will remain an important theme for years, not only in the United States but in every nation that aspires to participate in the world economy. The inherent dual-use nature of the most significant new technologies—the so-called convergent bio-, info-, and nanotechnologies—guarantees that the development of these fields and their underlying science will be accompanied by increasing concern for misuse. Concern for the misuse of specific substances that might be employed by terrorists took concrete form in a provision of the Public Health Security and Bioterrorism Preparedness and Response Act of 2002 (Public Law 107–188)—the so-called select agent rule—that requires registrations for institutions and clearances for individuals handling a list of pathogens and toxins. I believe the need for this kind of specific and restrictive legislation is infrequent and will be rare in the future.

The issue of how to respond to threats of bioterrorism has been an object of increased attention within the science and security communities since the deliberate contamination of U.S. mail with anthrax in October 2001. Well before that incident, the National Academy of Sciences had convened a committee, known as the Fink Committee, to consider responsible measures that might be taken to reduce the risk that advances in bioscience might be exploited for terrorism. The recent establishment by the Department of Health and Human Services of a National Science Advisory Board for Biosecurity (NSABB) completes a highly successful policy-making cycle that began with the report of this committee. This process is a model for future productive dialogue between science and government, and I appreciate the spirit of cooperation that all parties have exhibited during this period.

The visa issues to which I devoted last year's talk to this forum are still with us, and they are still serious. Fears that the newly introduced foreign student tracking system SEVIS would seriously impede the fall 2003 enrollment process did not materialize thanks to concerted efforts by the Department of Homeland Security and the educational institutions. But reports from key institutions indicate that foreign graduate

150 SCIENCE POLICY UP CLOSE

student applications are showing weakness, and serious obstacles remain for foreign scientists attending conferences and research activities in the United States. These issues are receiving attention at the highest levels of government because it is well understood that a healthy scientific enterprise is a global one. The United States has benefited substantially from a steady influx of talent and ideas from around the world, and we desire to continue that enrichment consistent with responsible security. Progress on visas must be deliberate, because it must not come at the expense of security. I expect this to be a continuing issue for several years.

Societal Impacts of Science and Technology

Concern about environmental and health effects of the technologies we employ in daily life is another important theme affecting the interaction of science and government. These are more familiar themes than terrorism, and despite the sharpness of public debate surrounding them, our system of societal communication and response is well equipped to manage the issues they present. The rapidly expanding capabilities of the Internet and wireless communications provide many opportunities for the public to learn about issues, form interest groups, and act to support their views. In our era, emerging societal concerns are unusually well aired and highly visible to policy makers and elected officials.

An example OSTP is following closely is the societal impact of nanotechnology. We worked with Congress to clarify concerns prior to the passage of the 21st Century Nanotechnology Research and Development Act that President Bush signed last December, and are working with agencies to ensure coordinated action. Research in nanotechnology is a priority in this administration, and agency programs in this area are coordinated through an office reporting to OSTP. The Nanotechnology Act includes a number of provisions related to societal concerns, including (1) that a research program be established on these issues, (2) that societal and ethical issues be integrated into all centers established by the program, and (3) that public input and outreach be integrated into the program. The bill further requires two studies by the National Research Council, one of which is on the responsible development of nanotechnology. Finally, the bill requires a center focused on societal and ethical issues of nanotechnology. PCAST is preparing itself to serve as the presidentially designated group required by the act to ensure these issues receive attention. This may seem heavy machinery for a problem that many scientists feel does not yet exist. The point is to act quickly to establish credible approaches to identifying and dealing with potential impacts of nanotechnology to preserve public credibility for this important emerging field.

Other environmental and health issues are even more visible, and more controversial, and many have ethical, economic, and international dimensions that go far beyond science. Global change, reproductive technology, and health impacts of chemicals in the environment fall in this category. These are exceptionally important issues, and how they are dealt with now will have long-term consequences for our nation and for the world. I take very seriously the recent statement signed by more than five dozen eminent scientists expressing concern that—to put it in my own words—science in these controversial areas might be undermined by politically motivated actions. This should always be a concern of government, as well as of scientists, and throughout history special arrangements have been made to protect the integrity of the scientific process. Not least of these arrangements was the establishment, nearly a century and a half ago, of the National Academy of Sciences, which provides the "gold standard" for technical advice. The National Academies and the panels they form through the National Research Council have been employed frequently by this administration.

This is a good occasion for me to state clearly that President Bush believes policy should be made with the best and most complete information possible and expects his administration to conduct its business with integrity and in a way that fulfills that belief. As director of the Office of Science and Technology Policy, I accept the responsibility to hear and respond to issues affecting the integrity of technical advice within this administration.

Some of the issues raised by the eminent scientists, and in other documents, media reports, and websites, have themselves become subjects of a controversy which I think it is in the best interest of science to get behind us. I cannot guarantee that each of the multitude of daily decisions upon which science policy depends will be made wisely or efficiently, but I can assure you that there is no intention by this administration to undermine or distort the products of that machinery. My office works with many organizations in science, engineering, higher education, and industry to identify and resolve problems that affect the science and technology enterprise. We are most effective when we have an opportunity to bring parties together to resolve mutual differences and devise corrective measures. I will continue to use my office to fulfill the president's expectations for scientific integrity to the best of my ability.

Meanwhile, some perspective is needed here. We have important work to do, serious challenges to meet, great opportunities to exploit. The intricate machinery of American science is the envy of the world because it works exceptionally well. It does so because the interface between government and the scientific community is broad and robust and remarkably apolitical. It is important to keep it that way.

Priority Highlights

Science policy entails more than setting budgets, but that is a major bottom line of the policy process. I do not have time to review the status of every priority that I have mentioned above, but some highlights are important to capture the flavor of science in this administration.

Health Sciences.   Funding during these four years to NIH has increased more than 40 percent, to $28.6 billion. In response to this unprecedented national commitment, NIH as a whole has adopted an important new roadmap for transforming new knowledge from its research programs into tangible benefits for society. Emerging interdisciplinary issues such as nutrition and aging, together with revolutionary capabilities for understanding the molecular origins of disease, health, and biological function, will continue to drive change within NIH.

PRESIDENTIAL SCIENCE ADVISOR I — 153

National Science Foundation. In four years the NSF budget has increased 30 percent over fiscal year 2001 to $5.7 billion. Much of this funding has gone to enhance the physical sciences and mathematics programs, where advances often provide the foundation for achievements in other areas, as well as increases to the social sciences and to the NSF education programs.

NASA. NASA has increased 13 percent, largely for exploration science that will spur new discoveries, enhance technology development, and excite the next generation of scientists and engineers. I will say more about the president's new vision for space exploration in a moment.

Department of Energy. Science and technology programs have increased 10 percent, in such important areas as basic physical science and advanced computing. As the agency sponsoring the largest share of physical science, DOE's Office of Science is increasingly viewed as a high leverage area for investment. DOE has engaged in years of intense planning, culminating recently in a multiyear facilities roadmap that assigns specific priorities to a spectrum of new projects.

Energy and environment. This administration is investing heavily in technologies for producing and using energy in environmentally friendly ways, from shorter-term demonstration projects for carbon-free power plants to the very long-term promise of nuclear fusion for clean, scalable power generation. In the intermediate term, technologies associated with the use of hydrogen as a medium for energy transport and storage are receiving a great deal of attention, not only in the United States but internationally. The president's Hydrogen Fuel Initiative is a $1.2 billion, five-year program aimed at developing the fuel cell and hydrogen infrastructure technologies needed to make pollution-free hydrogen fuel cell cars widely available by 2020.

Economic vitality. This administration has also, of course, launched initiatives directly related to the president's priority for economic vitality.

The president's tax relief plan includes making the research and experimentation tax credit permanent, thereby spurring the sustained, long-term investment in R&D. The president has also signed an executive order making manufacturing-related R&D a priority in two complementary federal grant programs that target small business innovation. The president's initiatives on nanotechnology and information technology have created strong incentives for private-sector R&D funding in these areas and have provided strong intellectual property protections to stimulate innovation and enhance U.S. competitiveness.

Space science and exploration. The president has committed the United States to a long-term human and robotic program to explore the solar system and beyond. Described by the president as "a journey, not a race," this plan differs profoundly from the Apollo paradigm of a single massive project requiring a budget spike and an aggressive schedule. The new vision is sustainable and long term, balancing robotic and human roles and using a step-by-step approach to address the risks and costs within a steady and realistic flow of resources. The vision also focuses on technology advances that are equally important to progress at home on Earth. Anticipated advances in robotics, human-computer interface, electronic and mechanical miniaturization, and applications of nanotechnology should continue the impressive record of space technology developments that benefit all Americans. Experience with other space programs has shown that a strong, sustained vision for space exploration, with clear and challenging milestones, will inspire future generations of young people to study math, science, and engineering.

I have emphasized the strengths of our research enterprise more than its weaknesses, because the strengths dominate. The weaknesses are also well documented and are receiving a great deal of attention. Among them are our continued reliance on imported intellectual talent for advanced work in science and engineering—a practice that is threatened by changes in our visa practices after 9/11. Improving the visa process by itself, however, will not solve this problem. Deep changes are required to improve educational practice and encourage wider participation in technical fields among underrepresented populations. Another challenge is the shrinking of horizons in industrial research, combined with post-Cold War stagnation in funding for research in physical sciences and engineering. These trends have combined to produce gaps in fields that are important for future technologies. PCAST and other advisory boards have examined these issues and recommended courses of action that are reflected today in national budget priorities. There are no magic bullets for these issues, however, and they will not be resolved in a single budget cycle, especially during a time of serious budget constraints.

The United States is investing more in research and development than all other G-8 nations combined. Current priorities for research funds clearly identify fields likely to be important for future economic competitiveness. The quality of research produced by our universities and industrial and national laboratories is unsurpassed by any other nation. As other nations develop their research capabilities and seek ways to reap economic payoffs from research investments, they emulate our structures and processes as best they can. As we act to make our system even stronger, let us be proud of the strengths of the United States research and development enterprise.

5

Presidential Science Advisor II

Measuring and Prioritizing

John Marburger remained presidential science advisor and director of the Office of Science and Technology Policy (OSTP) after George W. Bush was sworn in for his second term in January 2005. Over the next four years, Marburger's thoughts about science policy deepened and acquired greater scope. This was due to his greater experience, to his expanded role, and to his readings on science policy. As can be seen in the following addresses, he was particularly influenced by certain works about science policy by Daniel Sarewitz, professor of science and society at Arizona State University; Roger Pielke Jr. of the Center for Science and Technology Policy Research at the University of Colorado; and Frank Press, president of the U.S. National Academy of Sciences from 1981 to 1993. Marburger liked to say that his favorite science policy paper—"one that should be required reading for anyone interested in federal science policy"—was Sarewitz's "Does Science Policy Matter?"[1] As Marburger remarks in his final AAAS policy forum ad-

---

[1]Daniel Sarewitz, "Does Science Policy Matter?" *Issues in Science and Technology* (Summer 2007), http://issues.org/23-4/sarewitz/. Marburger first heard it as a talk in 2003. Despite the "decentralization of influence over S&T budgeting in the federal government" that "precludes any strategic approach to priority setting and funding allocations," Sarewitz writes, nevertheless, "one of the most astonishing aspects of science policy over the past 30 or so years is the consistency of R&D funding levels as a portion of the discretionary budget."

dress below, this paper crystallized his appreciation for two facts about science policy. One was that federal R&D funding has stayed virtually unchanged for years regardless of the administration or party controlling the White House. The second is that the fragmentation of science and technology policy implementation among so many uncoordinated actors makes it hard not just to realize the intent of a policy in practical terms but even to discover whatever impact a policy has. Marburger derived a twofold lesson from these facts. The most effective policy task that a presidential science advisor could undertake, he concluded, is, first, not to try to buck the historical trend and battle for an increase in the historically stable flow of money into science, but to try to make the infrastructure more efficient and effective (as Sarewitz had remarked in the second sentence of "Does Science Policy Matter?," "It is not only axiomatic but also true that federal science policy is largely played out as federal science budget policy"), and second, to foster development of better measures of the impact of science policy. During Bush's second term, Marburger was actively involved in major projects to address these two issues: the American Competitiveness Initiative and his promotion of what he called the "science of science policy."

—*RPC*

American Association for the Advancement of Science
30th Forum on Science and Technology Policy
Keynote Address no. 4
Washington, DC
April 21, 2005

Marburger opens this address by asserting his intent to focus "squarely on budgets and the measures of the strength of American science and technology." Robust science policy, he argues, requires not only knowing the budgets but also being able to "benchmark" R&D investments and their results. Again, this address is in background context of widespread concerns in the scientific community about Bush's commitment to science and technology. Just as in his second AAAS policy address he confronted criticism of the security-related restrictions of the scientific workforce with facts about the visa process, so here he counters critics of the Bush administration's science policy with budgetary facts. Marburger walks the audience through subtleties in the numbers, such as the difference between presidential budget requests and enacted budgets, and the often irrelevant distinction between basic and applied research. He cites Sarewitz's exposure as "myth" the idea that "the nation's commitment to basic research is weak," and Pielke to the effect that "over the past decade S&T has experienced a second golden age." Marburger then raises the question of how we know whether the current science and technology budget is appropriate for today's world. "I do not know of any reliable way to answer this question short of developing a massive econometric model for the world's economies and workforces, and exercising it with various scenarios." We have only a primitive framework, he says, "to evaluate policies and assess strength in science and technology." He cites a few "advocacy benchmarks" and then explains why these are useless for policy making and how they have been misused by journalists and commentators. "What science policy needs," he continues, "is the kind of quantitative tools economic policy makers have available, including a rich variety of econometric models, and a base of academic research." Then he broaches the need for "a new interdisciplinary field of quantitative science policy studies," which he will shortly call a "science of science policy." A month after this address, Marburger published a brief editorial in *Science* magazine that draws from this address. It concludes:

> Relating R&D to innovation in any but a general way is a tall order, but not a hopeless one. We need econometric models that encompass enough variables in a sufficient number of countries to produce reasonable simulations of the effect of specific policy choices. This need won't be satisfied by a few grants or workshops, but demands the attention of a specialist scholarly community. As more economists and social scientists turn to these issues, the effectiveness of science will grow, and of science advocacy too.[2]

The main concern for science policy, Marburger concludes, is not budgets, facilities, and restrictions on policies, but tools. This marks the beginning of a push for better science policy tools that he will keep up for the next several years.

—*RPC*

---

[2]John H. Marburger III, "Wanted: Better Benchmarks," *Science* 308 (May 20, 2005), 1087.

Thanks to the AAAS once again for organizing this annual event. While budgets are not the only thing on the agenda, the timing of this forum makes it clear that the top issue is the president's proposal to Congress for R&D spending in the forthcoming fiscal year. So I was surprised when I looked back at my remarks at three previous forums to find that I said relatively little about the details of President Bush's proposals, and more about the factors that lay behind them. Today I am going to focus squarely on budgets and the measures of the strength of American science and technology.

The sequence of R&D budgets during President Bush's administration very clearly shows a strong commitment to science and technology. Anyone looking at the graph below [Figure 1] can see that R&D growth in this administration is exceeded only by the buildup of federal funding in the post-*Sputnik* era of the early 1960s. This remarkable record has been parsed half to death by commentators, but its underlying message is unmistakable: this president and the Congresses that have worked with him regard strong federal R&D spending as essential to the health, security, and prosperity of the nation.

Part of my talk today is about this record, and part is about the rapidly changing context for R&D and what we need to do to make sense of

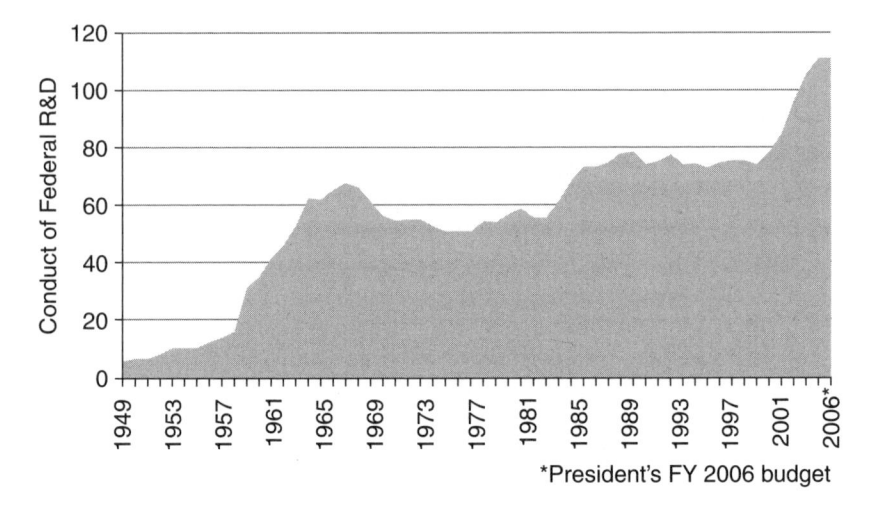

Figure 1. Federal R&D spending (outlays in billions, constant 2000 dollars).

it—to "benchmark" it, if you will. But first, the fiscal year 2006 budget and its history.

This year's budget is under considerable pressure. It maintains a strong focus on winning the war against terrorism while seriously moderating the growth in overall spending. Consequently, the fiscal year 2006 proposal is the tightest in nearly two decades.

Despite these pressures, federal R&D funding is actually *increased* in the president's request. And the administration has maintained high levels of support for the priority areas of nanotechnology, information technology, climate change science, and energy technology—including the hydrogen initiative—and space exploration. In a budget that would cut the total of "nonsecurity" discretionary spending by 1 percent from the 2005 allocated amount, total "nonsecurity" R&D spending is spared.

What this means is that the fiscal year 2006 proposal preserves the substantial increases in R&D spending made during the first term of this administration. The U.S. research and development enterprise is currently working from a new historically high base as it enters an era of rapidly changing conditions in global technical activity. Let me remind you of the actual numbers.

The president's fiscal year 2006 budget increases total R&D investment by $733 million to a new high of $132.3 billion, which is 45 percent greater than fiscal year 2001's $91.3 billion. The budget allocates 13.6 percent of total discretionary outlays to R&D—the highest level in thirty-seven years. Nondefense R&D accounts for 5.6 percent of total discretionary outlays, an amount significantly greater than the 5.0 percent average over the past three decades.

Some commentators have noticed that I have responded to concerns about the modest current growth rate in nondefense R&D by pointing to the enormous growth since fiscal year 2001. I do that because this investment has a real impact on the technically intensive sector of the American economy. The significance of such historic growth, however, is not acknowledged in widely publicized advocacy analyses of the health of the U.S. science and engineering enterprise. I will say more about those analyses in a moment, but let me point out now that they depend heavily on the NSF Science and Engineering Indicators for 2004, which are nearly all based on data collected through fiscal year 2001. These indicators measure the effect of the prior decade where R&D spending was indeed flat. They do not reflect the stimulus of the substantial correction in R&D budgets that actually occurred in the first term of the Bush administration.

Commentators also point to the large component of development expenditures—the "D" in R&D—in the R&D run-up of 2001–2005. In 1995 an important National Research Council committee chaired by Frank Press concluded that a more accurate measure of the investment in "the creation of new knowledge and the development of new technologies" would omit the "D" component.[3] That report is the origin of the budget category of federal science and technology (FS&T) first implemented in its present form in President Bush's 2002 budget proposal, but estimating its value back to 2000. It, too, increased substantially—30.4 percent—during the Great Advance from fiscal year 2001 to fiscal year 2005. This category has a short history, but I believe similar information is conveyed in the nondefense component of R&D shown below [Figure 2]. While I am uneasy about disregarding the "D" category altogether when we assess the portfolio of federal investments needed to keep our technology-based economy strong, I agree that FS&T is a better measure of long-term S&T investments.

The fiscal year 2006 request for the FS&T budget is $61 billion, a 1 percent reduction from the fiscal year 2005 enacted level. This is a good place for me to point out that presidential requests and prior year enacted budgets are not comparable because the enacted budgets include many congressionally directed programs (so-called "earmarks") that are not contained in the president's request. Enacted-to-enacted comparisons are valid; enacted-to-requested are not. The slight FS&T budget

---

[3]National Academy of Sciences, Committee on Criteria for Federal Support of Research and Development, *Allocating Federal Funds for Science and Technology* (Washington, DC: National Academies Press, 1995), www.nap.edu/openbook.php?record_id=5040 &page=R1.

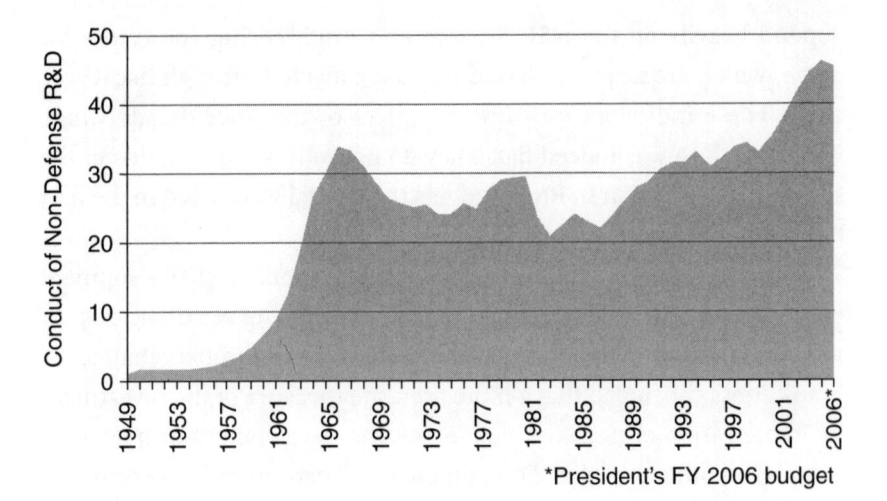

*President's FY 2006 budget

Figure 2. Nondefense federal R&D spending (outlays in billions, constant 2000 dollars).

decrease is entirely attributable to this mismatch. Earmarks in this portion of the budget exceed $2 billion. I would like to find a way to integrate congressional program direction with the executive branch planning and prioritization to optimize the use of federal funds for research.

Despite the strong recommendation of the 1995 Press committee, the old categories of basic and applied research continue to haunt some advocacy assessments of national S&T strength. In my opinion, this leads to seriously misleading conclusions. Hear these words from the Press report: "The committee's definition of FS&T deliberately blurs any distinction between basic and applied science or between science and technology. A complex relationship has evolved between basic and applied science and technology. In most instances, the linear sequential view of innovation is simplistic and misleading. Basic and applied science and technology are treated here as one interrelated enterprise, as they are conducted in the science and engineering schools of our universities and in federal laboratories." Ten years later the "complex relationship" has evolved to significantly new modes of research that are even more difficult to sort out among the old categories. The "basic research" category is nevertheless still tracked somehow by OMB, and it increased 26.2 percent during the Great Advance and stands at $26.6 billion in the fiscal year 2006 request, very slightly down from the $26.9 billion enacted level the prior year, the reduction once again due entirely to accounting for earmarks.

I want to underscore the significance of the recent history of R&D funding. Dan Sarewitz, a familiar science policy figure now at Arizona State University, has pointed out that "science policy discourse has been in the grip of a number of myths that seem utterly insensitive to the reality of [the] budgetary history. The first is that the nation's commitment to basic research is weak, and that basic science has been under continual assault by politicians who don't understand its value." Roger Pielke Jr., director of the Center for Science and Technology Policy Research at the University of Colorado, expressed some of my own frustration when he wrote recently that "few seem to be aware that over the past decade S&T has experienced a second golden age, at least as measured by federal funding, which has increased dramatically in recent years at a pace not seen since the 1960s."

We may still legitimately ask whether even these historically large amounts of R&D funding are right for the times. Questions like this are invariably raised in an international context. Are we funding all the R&D we need to defend ourselves, improve and sustain our quality of life, and compete with other nations in a globalized high-technology economy? I do not know of any reliable way to answer this question short of developing a massive econometric model for the world's economies and workforces, and exercising it with various scenarios. Two decades ago such a project would have seemed impossible. Today with modern information technology and the Internet I can imagine how it might be done. But we do not have such models now.

It is well to keep in mind how primitive the framework is that we use to evaluate policies and assess strength in science and technology. In the absence of models that link inputs like federal R&D investments to outputs like gross domestic product per capita, we collect annual data and fit straight lines to it to forecast future conditions. We try to interpret the data by taking various ratios, plotting the results in different ways—on semilog graphs, for example—and then talking about the results based on our intuitions about what it all means. Some of the results of this approach are useful for advocacy. They wake us up to changes so rapid they have to be important somehow—the rate of production of engineering degrees in China, for example, or rates of publication in technical journals, or government investments in different fields.

But let us not kid ourselves that these "benchmarks" contain information useful for policy making. Take the commonly quoted plot of federally funded R&D per unit of GDP. It has been going down in the United States for decades even as R&D funding has been going up. It has been going down on the average for OECD [Organization for Economic Cooperation and Development] nations for decades, and everywhere for the same reason: industry is doing more R&D all the time, and that is almost certainly related to why the GDP is going up so steadily in these countries. It is not bad for industry to be funding more research relative to the government, especially given the evolution that Frank Press's committee talked about a decade ago: basic and applied work are strongly merged in many important fields, and industrial R&D is adding significantly to the intellectual property base that supports important national objectives. The only major economy in which this ratio is going up (slowly) is Japan's, where nearly all of the R&D investment had been in the private sector, and Japan is finally adopting policies more similar to other developed nations.

Because of huge differences in how R&D is funded in different countries, it is better to compare the sum of public and private funding per GDP. I do not see any deep rationale for this ratio, especially in comparing economies of vastly different size, but it is the measure used by the OECD and other sources. (There is a good discussion of this ratio in the 2004 NSF Science and Engineering Indicators report.) This measure is much more stable than the ratio of government R&D alone to GDP and is used as a planning target within the European Union. The E.U. would like its members to spend 3 percent of GDP on R&D, but worldwide only two countries with large economies even come close: the U.S. with 2.7 percent and Japan with 3.3 percent, in both cases rising. In the United States, private funding is twice government funding.

Japan's ratio is converging to this, but U.S. government funding for R&D still exceeds Japan's in absolute terms by a factor of 3.

The misuse of ratios in widely publicized advocacy benchmarks seems to have misled some journalists and commentators. I read an article recently that claimed "the U.S. scientific enterprise is riddled with evidence that Americans have lost sight of the value of nonapplied, curiosity-driven research." Apart from the point that current ideas about research metrics tend to blur the distinction between pure and applied research, this statement is sharply contradicted by the recent history of funding in the basic research category. Total basic research expenditures during the past five years exceed those of the prior five years by 33 percent in constant dollars.

Although it is not useful for international comparisons, it is worth keeping in mind that the government portion of R&D has been a practically constant fraction of the U.S. domestic discretionary budget for decades. That is, more money goes to science in direct proportion to the money "on the table" during any budget year. The ratio is even more stable, at about 11 percent, if defense spending is excluded. This fact is like Moore's law—there is no necessity for nondefense science to receive about 11 percent of the nondefense discretionary budget year after year for decades, but it is happening, and it is a reasonable bet that it will continue to happen. This undermines arguments about particular influences on the top-line federal research budget to such an extent that Daniel Sarewitz has asked whether science policy even matters. Of course it does, because it is not just the top line that matters. Science policy plays itself out in the establishment and implementation of priorities within the available budgets. In times like the present when the discretionary budget is constrained, it is normal to find decreases as well as increases within the overall science portion of the budget.

The fiscal year 2006 R&D request highlights priority areas including some, like nanotechnology, that are often mentioned in international comparisons. The U.S. National Nanotechnology Initiative is a well-organized interdisciplinary program that has received much attention from Congress as well as the administration and benefits from a current investment of more than $1 billion across more than a dozen agencies. This budget has doubled within the past five years. During the past six months, the President's Council of Advisors on Science and Technology has looked in depth at the strength of the U.S. nanotechnology effort relative to other nations. PCAST found that, while the public sector investment (which includes not only federal expenditures but also state funding) in the United States is approximately equal to the investments by Europe and Asia, the United States leads the world in nanotechnology as measured by a number of different metrics, such as the number of scientific papers published and the number of patents filed. The PCAST report can be found through the OSTP website.

Another priority area that has received much public comment is math and science education. The president's fiscal year 2006 proposal requests an increase of $71 million, or 28 percent for the K-12 Math and Science Partnership program, an initiative designed to recruit postsecondary institutions to enrich math and science curricula in school districts throughout the country. This initiative is carried forward jointly by the National Science Foundation and the Department of Education, and as the program matures, funding has shifted between the two agencies. Reductions in the proposed budget for NSF are more than matched by requested increases for the portion of the program in the Department of Education.

The president's annual budget proposal to Congress is complex but assembled in a well-defined process. It reflects priorities that are explained in the budget narrative, which is available online. Some commentators and journalists work hard to understand its intricacies, and I strongly recommend that anyone interested in science funding regard this document as a primary source and read the science narrative carefully. As complex as it is, it is easier to understand the federal budget than it is to build an econometric model of the R&D enterprise.

Now I would like to return to that vision. Under the auspices of the National Science Board, the NSF Science and Engineering Indicators program produces an outstanding series whose volumes are full of analysis as well as data. Just as I urge you to read the president's budget proposal each year, I strongly suggest that you read as much of the narrative volume of the indicators as you can. Do not simply surf the statistical volume for numbers. Read what the text says about the numbers. This is an objective, high-quality document full of excellent insights.

That said, the indicators are based on a data taxonomy that is nearly three decades old. Methods for defining data in both public and private sectors are not well adapted to how R&D is actually conducted today. For example, all R&D carried out by a corporation is attributed to that corporation's main line of business. And the indicators are not linked to an overall interpretive framework that has been designed to inform policy. These problems and many more are analyzed in a very recent publication of the National Research Council titled "Measuring Research and Development Expenditures in the U.S. Economy" [2004]. On page 1 the authors write, "The NSF research and development expenditure data are often ill-suited for the purposes to which they have been employed. They attempt to quantify three traditional pieces of the R&D enterprise—basic research, applied research and development—when much of the engine of innovation stems from the intersection of these components, or in the details of each. . . . [T]he data are sometimes used to measure the output of R&D when in reality in measuring expenditures they reflect only one of the inputs to innovation and economic growth. It would be desirable to devise, test and, if possible, implement survey tools that more directly measure the economic output of R&D in terms of short-term and long-term innovation. Finally, the structure of the data collection is tied to models of the R&D performance that are increasingly unrepresentative of the whole of the R&D enterprise." The report makes a number of recommendations for improving various components of the data and enhancing their usefulness. These recommendations should receive high priority in future planning within NSF.

The growing importance of R&D within our society, however, and its strong association with national priorities, demands much more than the kind of improvements recommended in the NRC report. My perception of the field of science policy is that it is to a great extent a branch of economics, and its effective practice requires the kind of quantitative tools economic policy makers have available, including a rich variety of econometric models, and a base of academic research. Much of the available literature on science policy is being produced piecemeal by scientists who are experts in their fields but not necessarily in the methods and literature of the relevant social science disciplines needed to define appropriate data elements and create econometric models that can be useful to policy experts.

I am suggesting that the nascent field of the social science of science policy needs to grow up, and quickly, to provide a basis for understanding the enormously complex dynamic of today's global, technology-based society. We need models that can give us insight into the likely futures of the technical workforce and its response to different possible stimuli. We need models for the impact of globalization on technical work; for the impact of yet further revolutions in information technology on the work of scientists and engineers; for the effect on federal programs of the inexorable proliferation of research centers, institutes, and laboratories and their voracious appetite for federal funds; for the effect of huge fluctuations in state support for public universities. These are not items that you can just go out and buy, because research is necessary even to frame an approach. This is a task for a new interdisciplinary field of quantitative science policy studies.

I am confident about America's near-term future in science and technology, but I share the concerns of many about the longer term. I do not fear so much that our current budgets are too small, or that our facilities are inadequate, or that our policies guiding federal research are too restrictive. But I worry constantly that our tools for making wise decisions, and bringing along the American people and their elected representatives, are not yet sharp enough to manage the complexity of our evolving relationship with the awakening globe. I want to base advocacy on the best science we can muster to map our future in the world.

This annual forum sponsored by AAAS is an ideal place to stimulate interest in the work that needs to be done, and explain the relevance of policy studies to our nation's future. I congratulate the organizers of today's event on an excellent agenda. Thank you.

American Association for the Advancement of Science
31st Forum on Science and Technology Policy
Keynote Address no. 5
Washington, DC
April 20, 2006

In his State of the Union address of January 31, 2006, President George W. Bush announced the American Competitiveness Initiative (ACI). This act was largely Marburger's doing, though he never sought to take credit for it. As background, while he was chairman of the board of Universities Research Association (URA), the group managing the Superconducting Super Collider, Marburger heard a speech given by National Academy of Sciences president Frank Press in which Press proposed ways to prioritize scientific projects during a time of restricted resources. "The word *prioritize* sends shivers down the backs of most science advocates," Marburger remarks here, recalling the scientific community's horror and disbelief at the prospect. In the twenty years following Press's speech, however, as science projects continued to increase in scale and expense, and as constraints on the discretionary budget rose as well—despite a sharp rise in federal R&D funding during the first Bush administration—the need to prioritize had become even more urgent and obvious. Marburger also recalled the 1995 report of a committee chaired by Press pointing to economic competitiveness as "among the most important reasons for the investment of public funds in long term high risk investigations." The ACI, described in this address, was Marburger's response to that challenge, his attempt to establish a principled framework for the federal government to fund priorities among programs and agencies—"science in the national interest," though he did not use that phrase. The ACI involved an attempt to create a sustainable science funding platform over time; to identify what parts of scientific research it was appropriate for the federal government to fund as opposed to other sponsors; to establish an appropriate growth rate for underfunded prioritized areas, drawing on lessons learned from the recent attempt to double the NIH budget; and to establish educational initiatives. The ACI was the Bush administration's science budget policy from the time it was announced to the end of the administration, and what Congress would eventually authorize, under the name "America Creating Opportunities to Meaningfully

Promote Excellence in Technology, Education, and Science Act of 2007" (the America COMPETES Act), was modeled on the ACI and follows its numbers closely. Marburger's role in developing the ACI is an excellent illustration of his approach to science policy as presidential science advisor: he viewed his role as not simply to be an advocate for science but to press the case for how science could address the various problems that the administration was trying to solve.

—RPC

Thanks to AAAS for inviting me once again to address this important annual policy forum. The pace of scientific discoveries far exceeds the pace of science policy, so you would expect these annual forums to be rather repetitive and boring, especially when you hear from the same people—like the president's science advisor—year after year. This is the fifth time I have spoken in the forum, and I will try to avoid repeating myself, although I admit I am tempted. Instead, I will repeat the words of other science advisors, starting with Allan Bromley. In his book about his experience as science advisor to the first President Bush, Bromley said one of his most surprising discoveries about Washington was that "it took longer to make anything happen than I could have believed possible!" This fact of Washington life means that often years go by between the emergence of challenges and effective responses to them. And yet things do happen, and government can act decisively when the path forward is clear.

This year President Bush launched not one but two significant science-based initiatives in his State of the Union message. The president's subsequent vigorous promulgation of both initiatives in forums, tours, and speeches demonstrates the seriousness with which he regards these programs as part of his domestic policy agenda. Just yesterday I accompanied the president to Tuskegee University in Alabama where he toured an NSF- and DOD-sponsored nanomaterials laboratory and spoke about his initiatives in research and education. Today I want to put one of these programs, the American Competitiveness Initiative, in historical perspective. The other, the Advanced Energy Initiative, also deserves attention by this audience, but it has a smaller policy foot-

PRESIDENTIAL SCIENCE ADVISOR II

print. These are long-term initiatives based on a conviction that science is fundamentally important for our future economic competitiveness, and for national and homeland security, the president's three highest priorities. In the larger context of post-Cold War science policy, the American Competitiveness Initiative—which I will refer to as the ACI—is part of a long evolution that began in the early 1990s and will likely continue into the next decade. Before I speak of that history, I want to recall the immediate context for this year's science budget and remind you of the specific features of the ACI.

I know that even many policy wonks do not read the text of the annual budget proposal the president sends to Congress each year (it sits in four thick volumes on my bookshelf). But I strongly urge you to read at least the front matter in the budget—the president's budget message, the overview, and the articles on special issues. These sections signal what the administration will emphasize in the ensuing interactions with Congress to produce a final budget. I want particularly to stress the president's determination to reduce the federal budget deficit during his second term in office. In his own words: "Last year, I proposed to hold overall discretionary spending growth below the rate of inflation—and Congress delivered on that goal. Last year, I proposed that we focus our resources on defense and homeland security and cut elsewhere—and Congress delivered on that goal. . . . The 2007 Budget builds on these efforts. Again, I am proposing to hold overall discretionary spending below the rate of inflation and to cut spending in non-security discretionary programs below 2006 levels." In the same message, the president also said "And my Budget includes an American Competitiveness Initiative that targets funding to advance technology, better prepare American children in math and science, develop and train a high-tech workforce, and further strengthen the environment for private-sector innovation and entrepreneurship."

The question immediately comes to mind: how are you going to fund a major initiative like this and cut spending at the same time? The ACI proposes nearly a billion dollars ($910 million) of new research funding for three specific agencies, and a commitment to double their combined budgets over ten years—a total of $50 billion during that period. To increase incentives for industrial research, the budget would forego $4.6 billion of tax receipts for companies that invest in research and experimentation, for a ten-year cost of $86.4 billion. The education component of the ACI would add another $380 million in fiscal year 2007. The president's fiscal year 2007 budget does not increase overall nondefense discretionary budget authority, and the request for nondefense R&D budget is proposed to increase at a rate slightly less than inflation. Accommodating the ACI in a flat or declining budget is only possible by setting priorities and allocating funds differentially to the highest priority programs. The key phrase in the president's reference to this program is the "American Competitiveness Initiative that *targets* funding." The fiscal year 2007 budget for science is very clearly about priorities.

The word *prioritize* sends shivers down the backs of most science advocates. Eighteen years ago this month, toward the end of the Reagan administration, National Academy of Sciences president Frank Press gave a memorable speech at the 125th annual meeting of the academy titled "The Dilemma of the Golden Age." It was a shocker of a speech because Frank made concrete proposals for how to prioritize science in a time of fiscal constraint and urged his colleagues to participate in the process. Some of you here today may recall that valiant effort. I looked up the subsequent press coverage in preparation for a speech I gave last year in a memorial symposium at Yale for Allan Bromley.

Here's Al Trivelpiece, then executive director of the American Association for the Advancement of Science: "Nobody asks farmers whether they want price supports for wheat rather than for cotton. Why should scientists be treated any differently and be required to choose from among several worthy projects? I think the issue for scientists should be the quality of the research." And a congressional staffer: "I hope we can forget his words and move on." And an official of the American Association of Medical Colleges: "It's a question of strategy. Why should we assume that there's a fixed pot of dollars? I prefer the idea that support for science is not fixed, at least not until we get to a level that represents a reasonable proportion of our GNP." Science advocates—which include the major associations and professional societies—distanced themselves from Press's suggestion that the science community itself is the best place to look for guidance in establishing priorities across fields and programs.

I regard Frank's 1988 address as a key document in the history of American science policy. He prepared it at a time when scientists were chafing under the crunch of a serious budget deficit that the president and Congress were struggling to get under control. Research opportunities were outstripping growth in the nondefense federal science budget, and different sectors of the science community were sniping nastily at each other. I recall it very well because at the time I was chairman of the board of Universities Research Association, then competing to build the SSC, which was the target of some of that sniping. Press explained the "dilemma" in his title as a product of the "very exuberance—in that golden age of discovery and advance. [And now I'm quoting extensively from his speech.] Our scientists are submitting in record numbers proposals of the highest quality, with enormous intellectual and material potential. We have also laid on the budget table very large and very expensive new ventures—in multiple fields from high-energy physics to molecular biology, whose time in the progress of science has arrived. The proposals—small and large—are superb in quality, but unprecedented in overall cost. And the reality is that these proposals come at a time of record budget deficits."

"There," said Frank, "is the heart of the dilemma. It is not the lack of political support for science. Political decision makers in the executive branch and Congress no longer need convincing that leadership of American science and technology is vital to our nation's future. The real political issue is what does science most urgently need to retain its strength and its excellence. The issues are funding levels and priorities."

These remain the issues today, but in the intervening years something very important has happened, and the atmosphere is different. I am not sure there ever was a time that scientists felt their sponsored funds were commensurate with their opportunities for discovery, and frustration over that gap is widespread to this day. But I no longer see the sharp-edged ill-will among different fields that worried Press nearly twenty years ago. Those were the final years of Cold War science policy and cracks had begun to appear in the framework of mutual understanding among sponsors and researchers that had supported science since *Sputnik*. Today we are emerging from that long transition in U.S. science policy I mentioned that began at about the time of Frank Press's 1988 address. Let me reflect for a moment on what happened during that period.

In 1989 the Berlin Wall came down, and Tim Berners-Lee and colleagues at CERN launched the World Wide Web. Two years later historians declared the Cold War officially over, and Congress began looking for a peace dividend. Within the Department of Defense, the largest sponsor of university-based engineering research, science was not spared. In 1993 Congress terminated the Superconducting Super Collider and narrowly authorized the International Space Station project with a margin of one vote. The ebbing tide of Cold War weapons production had revealed a huge problem of environmental contamination at Department of Energy weapons facilities, and DOE science funding went flat. House Science Committee chairman George Brown admonished scientists in general, and physical scientists in particular, to seek a new post-Cold War rationale for government funding of their work. Industries that had supported productive research laboratories began reducing budgets and shrinking their horizons. Some were reacting to reductions in defense spending, and others to deregulation and continued competitive pressure from Japan and the then emerging Asian "tiger economies."

Science, meanwhile, saw new horizons opening with the almost miraculous appearance of powerful tools generated by the information technology revolution. If Frank Press's late 1980s were a "golden age" for science, the 1990s revealed a platinum or even a diamond age of discovery based on new capabilities for managing complex or data-intensive systems, and especially the extraordinarily complex systems of the life sciences. The coming twenty-first century was described as the century of biology in contrast with the old century of physics. The new technologies, to be sure, were based on physical science, but it appeared to be a known and reliable physical science that had provided an inventory of capabilities "on the shelf" that the military or industry could exploit in its own new breed of shorter-horizon, development-oriented R&D laboratories. Industrial research made Moore's "law" come true during this decade and produced the devices and systems that lured entrepreneurs and their financial backers into the dotcom bubble. These conditions tended to obscure the role of basic research in the physical sciences and depress the perception of its importance in the agencies on which the field had depended since World War II.

The obvious changes in conditions for science during the 1990s stimulated a variety of interesting policy responses. Frank Press made another important contribution as chair of an NRC committee that produced the 1995 report "Allocating Federal Funds for Science and Technology."[4] This report summarized current thinking about the case for federal funding of research and added to a growing consensus that economic competitiveness was among the most important reasons for the investment of public funds in long-term, high-risk investigations. It was an important precursor to the subsequent report prepared by Congressman Vernon Ehlers at the request of Speaker Newt Gingrich in 1998. This document, *Unlocking Our Future: Toward a New National Science Policy*,[5] concluded that, "because the scientific enterprise is a critical driver of the Nation's economy, investment in basic scientific research is a long-term economic imperative. To maintain our Nation's economic strength and our international competitiveness, Congress should make stable and substantial federal funding for fundamental scientific research a high priority."

Neither of these reports addressed the problem of prioritization, but by the end of the 1990s it was clear that federal funding for biomedical research was racing ahead of funding for other fields. In actions widely regarded as demonstrating the increasing dependence of life science on physical science technology, Harold Varmus, the

---

[4]Ibid.

[5]U.S. House Committee on Science, *Unlocking Our Future: Toward a New National Science Policy*, Document 105-B (Washington, DC: U.S. Government Printing Office, 1998), www.gpo.gov/fdsys/pkg/GPO-CPRT-105hprt105-b/pdf/GPO-CPRT-105hprt105-b .pdf.

Clinton administration's NIH director, funded the construction of beam lines at the Department of Energy's x-ray synchrotron light sources and spoke eloquently of the importance of physics to biology and medicine. With the completion of the doubling of its budget over five years ending in 2003, NIH consumed roughly half the nondefense federal research funding, with NASA in second place with 15 percent. NASA's science budget alone has been comparable to the entire budget for NSF. DOE Office of Science budgets, which include funding for the powerful tools used to unravel the atomic structures of complex materials and biomolecules, have remained virtually flat. Concerns about the balance of funding surfaced explicitly in a document prepared by a subcommittee of PCAST chaired by Georgia Tech president Wayne Clough in 2002. The report, "Assessing the U.S. R&D Investment,"[6] stated, among other things, that "all evidence points to a need to improve funding levels for physical sciences and engineering." At the time, the country was still suffering the economic consequences of the burst dotcom bubble and was realigning budget priorities in response to the terrorist attacks the previous September. Completing an administration commitment to double the NIH budget was the highest science priority at that time, next to establishing an entirely new science and technology initiative for homeland security. Nevertheless, the administration continued to expand funding for targeted areas of physical science, including the recently introduced National Nanotechnology Initiative, and maintained funding for the Networking and Information Technology Research and Development program. The NSF budget continued to increase at a rate above inflation. In the first term of the Bush administration, combined federal R&D funding soared at a rate unmatched since the early years of the Apollo program, a jump of 45 percent in constant dollars over four years.

As the Bush administration concluded its first term, further reports began to appear that linked federal programs for research and educa-

---

[6]The President's Council of Advisors on Science and Technology, "Assessing the U.S. R&D Investment: Findings and Proposed Actions," www.whitehouse.gov/sites /default/files/microsites/ostp/pcast-02-rdinvestment.pdf.

PRESIDENTIAL SCIENCE ADVISOR II   177

tion to economic competitiveness, including two more from PCAST and one from the Council on Competitiveness in 2004. The following year, 2005, witnessed a growing wave of reports and publications with similar themes, culminating in the widely publicized report from the National Academy of Science *Rising Above the Gathering Storm: Energizing and Employing America for a Brighter Economic Future.*[7] These reports contributed to a clear basis for establishing funding priorities among programs and agencies in the ACI initiative, launched in President Bush's January 2006 State of the Union message. The policy principles are, first, that funding long-term, high-risk research is a federal responsibility; second, that areas of science most likely to contribute to long-term economic competitiveness should receive priority; and third, that current levels of funding for research in the physical sciences are too low in many agencies.

The American Competitiveness Initiative identifies the National Science Foundation, the Department of Energy Office of Science, the National Institute for Standards and Technology, and the Department of Defense as key agencies with major funding satisfying the three principles, and seeks to double the budgets of the first three over the next decade. The current year increase for the sum of the three is 9.3 percent. I described other components of the ACI earlier, and will not say more about them here. My point in recounting history since Frank Press's 1988 speech is to contrast the reluctance of nongovernment science stakeholders at that time to discuss priorities among different fields with what can be read as a consensus within some of the same communities today that even in a time of budgetary constraint something needs to be done with the budgets in at least some areas of physical science research.

The elements of the ACI resemble some of the recommendations made in the National Academies *Gathering Storm* report but is not

---

[7]National Academy of Sciences, National Academy of Engineering, and Institute of Medicine, *Rising Above the Gathering Storm: Energizing and Employing America for a Brighter Economic Future* (Washington, DC: National Academies Press, 2007), http://www.nap.edu/catalog.php?record_id=11463.

intended to be a direct implementation of those recommendations, many of which overlap existing federal programs or were expressed in a degree of generality incompatible with the kind of specificity required in a presidential budget proposal. There is no question, however, that the *Gathering Storm* report played an important role in bringing diverse components together under the theme of economic competitiveness and created an atmosphere in which such a complex set of proposals could receive favorable treatment by Congress. The report's authors, and particularly the committee chairman, Norman Augustine, deserve a great deal of credit for investing time and energy to raise awareness of the need for a set of coordinated actions to ensure the future economic competitiveness of our nation.

I am approaching the end of my talk, and I have said little about the budgets of other areas of science, or the details of how the ACI can be funded without serious negative impacts on other areas of science funding. The fact is that the fiscal year 2007 cost of the ACI is dwarfed by the $2.7 billion in current year earmarks in the research budget. Earmarking has increased rapidly during the past five years and has reached the point where it now threatens the missions of the agencies whose funds have been directed toward purposes that do not support the agency work plans. From the point of view of transparency in government operations, earmarking at this level erodes the value of reported budget numbers for inferring agency resources. For example, the $137 million in earmarks on the $570 million NIST core budget in the current year led to a gross exaggeration of how much money NIST actually has to satisfy its needs, particularly its physical plant requirements. The ACI request would increase the amount actually available to NIST by 24 percent, but because the earmarks mask the actual current amount, a comparison of the fiscal year 2007 request with the fiscal year 2006 appropriated suggests a *reduction* of 5.8 percent for NIST. This is a very serious problem. Media reporters attempting to identify "winners and losers" cannot even get the sign right on the budget changes inferred this way.

The White House Office of Management and Budget has criteria for identifying and accounting for earmarks, but those criteria are not employed by AAAS analysts, and the AAAS earmark methodology is not transparent. Unfortunately, OMB does not publish earmark data or include the effects of earmarks in its tables. Consequently, the dramatic growth of earmarks has seriously undermined the usefulness of the historically valuable OMB and AAAS analyses. Published budget numbers from either source no longer consistently reflect the actual resources available to science agencies to carry out their programs. This is not a satisfactory situation, and I urge AAAS to work with OSTP and OMB to develop a mutually comprehensible approach to the problem of taking earmarks into account in analyzing the annual science budgets.

Earmarking and prioritization are clearly related. One person's priority is another's earmark. One of the drivers for earmarking is the reluctance of individuals or institutions to participate in the merit-based review procedures that are best practices in most funding agencies today. Another is the absence of funding programs for categories of expense that are deemed important even sometimes by the targeted agencies. I believe that where science stakeholders can form a consensus on priorities, the negative impact of earmarking can be greatly diminished.

I wish to thank AAAS and its members for providing not only this but many other opportunities for bringing together the disparate sectors of the nation's science community, and working to build a consensus for constructive federal science policy.

American Association for the Advancement of Science
32nd Forum on Science and Technology Policy
Keynote Address no. 6
Washington, DC
April 2007

This speech is devoted largely to issues that Marburger had broached in previous forum addresses. Yet even when going over subjects that he has discussed at previous forums, such as the ACI, earmarks, priorities, and misleading aspects of familiar benchmarks, he is refreshingly articulate. For instance: "No law of nature or of politics guarantees that this real-life science posture [of the sum total of the myriad activities of a nation's "scientists and engineers, students and technicians"] will reflect a sensible scientific policy." At the end, moreover, Marburger introduces a new theme, having to do with his observation that the nation's biomedical research enterprise can be thought of "as a miniature economy with its own labor pool, markets, productive capacity, and business cycles." This development has resulted in a vast expansion of research capacity and is marked by diversified revenue sources, not just federal ones. As a result, "state and private sector resources should be considered more systematically in formulating federal science policy." Marburger concludes by calling on the AAAS forum to study this development with an eye to possibly helping to guide it in the future. One topic mentioned here in passing—in the simple sentence "climate change demands attention"—deserves further background. The subject was much in the news in 2007 and was another issue crystallizing criticism against the administration. *An Inconvenient Truth,* a movie by Al Gore, George W. Bush's opponent in the 2000 election, had premiered the previous year. Near the beginning of the movie, Gore relates a story about a teacher of his who humiliated a classmate in the course of offering an incorrect scientific explanation. "The teacher," Gore jokes, "went on to become science advisor in the current administration." Marburger wrote a reply to this nasty and demonstrably false swipe, never published, in which he pointed to the fact that the Bush administration had funded investigation into global warming as well as investments into energy technologies that potentially could be scaled to achieve reductions in greenhouse gas emissions. In this unpublished reply, Marburger quoted a 2001 remark by Bush that "we know the surface temperature of the earth is warming . . . and the National Academy of Sciences indicates that the increase is due in large part to human activity. . . . The policy challenge is to act in a serious and sensible way, given the limits of our knowledge. While scientific uncertainties remain, we can begin now to address the factors that contribute to climate change." Marburger also cited an exhibit—the climate change mitigation simulator, now online—at the National Academies' Marian Koshland Museum of Science in Washington (viewable at www.koshland-science-mu seum.org), whose point is the difficulty of adopting a policy that will make a significant impact on emissions. "The choices seem reasonable," Marburger writes, "but their impact is distressingly small. Nothing short of revolutionary changes in the world's energy technologies will make a real difference." Marburger also pointed to the failure of both the Gore film and the Koshland Museum to examine the politically unpopular option of nuclear power [this has now been incorporated into the online simulation –RPC]. Later that year, the Intergovernmental Panel on Climate Change (IPCC) would issue a report, much of which was based on science funded by the United States during the Bush administration, and negotiated at a 2007 conference with the OSTP as the lead U.S. delegation. At the end of the year, IPCC and Gore would share the Nobel Prize "for their efforts to build up and disseminate greater knowledge about man-made climate change, and to lay the foundations for the measures that are needed to counteract such change."

—RPC

Thanks to the AAAS for inviting me to speak once again at this annual policy forum. This is my sixth consecutive appearance as President Bush's science advisor and director of the White House Office of Science and Technology Policy. Before preparing my remarks I went back over the previous five speeches to make sure I was not repeating myself. Not that repeating is unwarranted. Every year our multiple science communities scrutinize so intently the individual budget trees of their respective fields that I am always concerned that they miss the forest of broader policy issues that affect them all. My intention over the years has been to draw attention to long-term issues and use whatever power my office has to address them apart from the frenzy of the annual budget cycle. Today I will repeat myself to remind you of some of those issues and add one more that deserves your attention. Along the way, I have a few observations about the current budget, too.

It is true that government-driven solutions to long-term problems require short-term authorizations, appropriations, and allocations in the budget process. And it is true that the budget process responds to many forces that pull it in different directions. It is all too easy for each of us to believe that the pressures we feel in our own work—research or its sustenance—can be solved by passage of a bill in Congress—usually one that adds funds that maintain our particular laboratories and research groups. The advocacy that we perform individually or through our institutions or professional societies shapes the actions of government, whose impacts spread throughout society. Ultimately, the science posture of a nation expresses itself in the myriad activities of its scientists and engineers, students and technicians—activities that may or may not sum to a coherent or effective whole. No law of nature or of politics guarantees that this real-life science *posture* will reflect a sensible science *policy*. The only hope of coherence in our national science posture is for all the diverse actors to agree on a general direction and give it priority year after year.

Such a consensus has been achieved on some important science policy issues during the past six years. Following the terrorist attacks of September 11, 2001, the science community came together in a remarkable show of unity to support what would obviously be a difficult and protracted struggle against terrorism. My AAAS policy forum speeches from 2002 to 2004 featured science and technology dimensions of antiterrorism, including the creation of a Science and Technology unit within the Department of Homeland Security, and a long list of initiatives to recruit science to the cause of homeland security. I also raised and reinforced concerns about the negative impacts of security measures on the conduct of science, and reported on actions OSTP and relevant departments and agencies were taking to mitigate these impacts. This is a continuing area of concern that deserves constant attention from the science community. While the student visa situation is much improved, we still have serious problems with a cumbersome and graceless visa process for visiting scientists, implementation of the export control regime, potential over-regulation of dual-use bioscience, and security arrangements that stifle user programs at key national laboratories.

The good news is that there *is* a consensus among nearly all actors that these are problems that need to be addressed. The danger is that with time the salience of these issues will diminish and momentum toward solutions will be lost. Within government, a number of interagency committees have sprung up to address problems at the intersection of science and security, and many other organizations, including the AAAS, have created committees and ongoing activities that will keep up the momentum. For example, the National Science Advisory Board for Biosecurity (NSABB) that I mentioned in my 2004 address has been meeting quarterly since the summer of 2005. For an impression of the care with which this group is considering the problem of "dual-use" bioscience, see the draft of the report it considered at its meeting last month, "Proposed Strategies for Minimizing the Potential Misuse of Life Sciences Research," available on the NSABB website. Another good example of agency responsiveness is the action of the Department of Commerce in reconsidering a proposed rule affecting the implementation of the export control regulations (ITAR [International Traffic in Arms Regulations]). The National Academies' Government-University-Industry Roundtable on this issue, chaired by Marye Anne Fox, has played a useful role in bringing parties together on this set of problems. I mention these issues today because I think they are not getting as much visibility as they deserve in the science and technology media.

Wide consensus also exists on the importance of federally funded science to our nation's long-term economic competitiveness. I spoke about the history of this consensus in last year's policy forum. The National Academies' 2005 report *Rising Above the Gathering Storm*[8] was an important expression of this view and echoed findings of many other reports. Notable among its recommendations was increased funding for basic research in the physical sciences, mathematics, and engineering—areas that had stagnated while the budget for biomedical research soared. The report even recommended that investment in these areas should increase "ideally through reallocation of existing funds, but if necessary via new funds." That statement is a rare recognition of the fact that federal funds for science are limited and that some programs may have to be held constant or reduced to fund priorities.

The administration's response to this consensus was the American Competitiveness Initiative, which among other things proposed doubling budgets for NSF, NIST and the Department of Energy's Office of

---

[8]Ibid.

Science over ten years. Appropriation committees in the 109th Congress produced bills that would have fully funded the ACI, but unfortunately, and to the great dismay of the very large number of ACI supporters, the Congress retired without passing the necessary bills. The previous Congress did, however, pass appropriations bills for the Departments of Defense and Homeland Security. I will come back to the implications of these bills in a moment. The new Congress adopted a continuing resolution (CR) that froze all discretionary budgets at fiscal year 2006 levels with two very important provisions: First, the 110th Congress used flexibility within the CR to fund the president's proposed ACI science budget for fiscal year 2007, but only at half the requested level. Second, it adopted a rule suspending earmark requirements on the continuing funds that had been earmarked in fiscal year 2006. This has a profound impact on the budget discussion for fiscal year 2008. Let me explain with an example.

The AAAS analysis of the president's fiscal year 2008 budget request states that "once again there would be steep cuts in DOD's S&T . . . programs. DOD S&T would plummet 20.1 percent down to $10.9 billion, with cuts in all three categories of basic research, applied research, and technology development." The figure below [Figure 3], taken from the AAAS report, shows these dramatic figures in dark gray.

As I explained in my remarks last year, "The fact is that the . . . cost of the ACI is dwarfed by the $2.7 billion in current year earmarks in the research budget. Earmarking has increased rapidly during the past five years and has reached the point where it now threatens the missions of the agencies whose funds have been directed toward purposes that do not support the agency work plans. From the point of view of transparency in government operations, earmarking at this level erodes the value of reported budget numbers for inferring agency resources. . . . This is a very serious problem. Media reporters attempting to identify 'winners and losers' cannot even get the sign right on the budget changes inferred this way." That's what I said last year. What readers of the AAAS report need to know is that the entire change in the fiscal year 2008 presidential budget request for DOD S&T comes from removing the fiscal year 2007 earmarks to determine a meaningful base

PRESIDENTIAL SCIENCE ADVISOR II   185

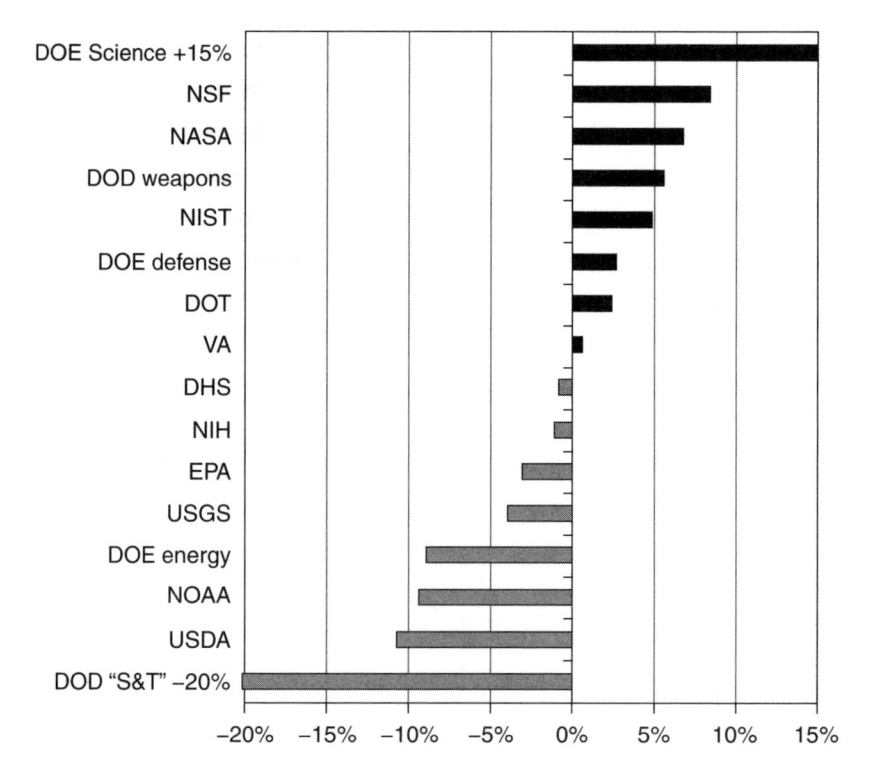

Figure 3. Fiscal year 2008 R&D request (revised): percent change from fiscal year 2007. *Source:* AAAS (March 2007 revised), based on OMB R&D data and agency estimates for fiscal year 2008. DOD "S&T" = DOD R&D in "6.1" through "6.3" categories plus medical research. DOD weapons = DOD R&D in "6.4" and higher categories. Fiscal year 2007 and 2008 figures include requested supplementals. FY 2007 = latest estimates of final appropriations.

budget for this important research. The president is actually asking Congress to *increase* the S&T budget that DOD can devote to its core programs, as shown in the second figure [Figure 4], which removes the DOD earmarks.

This failure by the AAAS to explain the treatment of earmarks in the administration's budget proposals is not good. It is a serious and unacceptable flaw in a report that is widely used as an authoritative reference on the budget. I am particularly disappointed by this lapse because last year at this time I pleaded that "the White House Office of Management and Budget has criteria for identifying and accounting

186   SCIENCE POLICY UP CLOSE

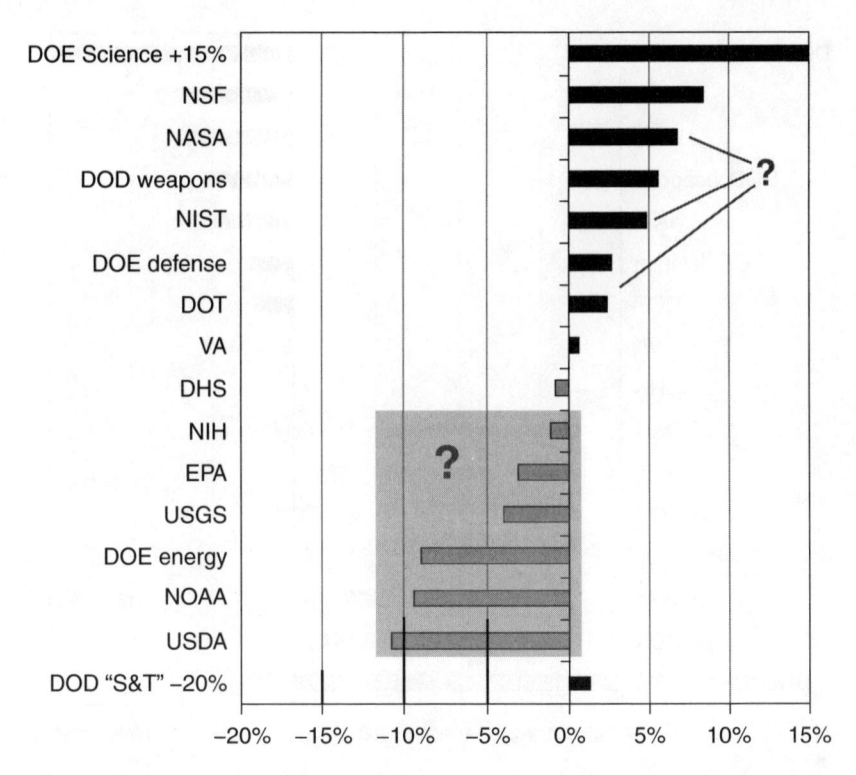

Figure 4. Fiscal year 2008 R&D request (revised): percent change from fiscal
year 2007. *Source:* AAAS (March 2007 revised), based on OMB R&D data and agency
estimates for fiscal year 2008. DOD "S&T" = DOD R&D in "6.1 through "6.3"
categories plus medical research. DOD weapons = DOD R&D in "6.4" and higher
categories. Fiscal year 2007 and 2008 figures include requested supplementals.
FY 2007 = latest estimates of final appropriations.

for earmarks, but those criteria are not employed by AAAS analysis,
and the AAAS earmark methodology is not transparent. Unfortunately,
OMB does not publish earmark data or include the effects of earmarks
in its tables. Consequently, the dramatic growth of earmarks has seriously undermined the usefulness of the historically valuable OMB and
AAAS analyses. Published budget numbers from either source no
longer consistently reflect the actual resources available to science
agencies to carry out their programs. This is not a satisfactory situation, and I urge AAAS to work with OSTP and OMB to develop a
mutually comprehensible approach to the problem of taking earmarks

into account in analyzing the annual science budgets." End quote from last year.

Well, AAAS has done nothing to correct its practice, but OMB has. Now you can look on the OMB website (http://earmarks.omb.gov) to browse the 13,497 earmarks that occurred in 2005 by agency or by state. In that year the earmarks totaled almost $19 billion, more than half of which was in DOD alone.

This year is an especially important one for understanding the status of earmarks because Congress, to its great credit, has passed on the prior year's earmarked funds to the agencies *without the requirement that they direct the funds to the earmarked purposes.* That is, they can—if they accept this generosity at face value and choose to ignore the original restrictive language—add the funds to programs that have been planned, prioritized, and properly evaluated to satisfy the mission needs of their departments and agencies. This is a very remarkable gesture that has effectively given agencies a huge windfall for the current fiscal year 2007 budget year. This action could not be taken into account in the president's budget proposal, which had to be prepared before Congress made its decision to remove the earmarking restrictions from the continuing resolution. I emphasize that this action did not apply to the hugely earmarked DOD budget, which is not subject to the CR.

What happens next will be extremely interesting. If Congress permits earmarks in its fiscal year 2008 appropriations bills, it will in effect be taking away the agency flexibility it granted in the continuing resolution, returning budgets the agencies can evaluate and use effectively to the base the president uses in his requests. President Bush has asked Congress to cut the total amount of earmarks in half. If Congress does that for the science budgets—without removing the associated funds it granted in the CR—it would be wonderful for science.

What Congress decides to do here will signal its priorities for research. The ACI prioritizes basic research in key agencies that have been relatively underfunded given the importance of the fields they support for long-term economic competitiveness. Because two Congresses have now failed to fund the first year of ACI at the level the president has requested, it is now behind schedule. The administration's fiscal year 2008 request aims to catch up. The administration continues to believe it is essential to rectify a long-growing imbalance in the pattern of research funding affecting the prioritized agencies. Despite much good will toward the ACI, and recent actions on competitiveness bills by authorizing committees in both the House and the Senate, the fate of this important initiative remains in doubt. What these agencies need is appropriations for their underfunded basic research programs. They do not need new programs or new bureaucracy, new reporting requirements, or new constraints on how they use their funds, all of which are features of the authorization bills. My plea to Congress is that it protect the basic research aims of the ACI from suffocation under the weight of all these other trimmings—twenty new programs in the Senate bill alone.

Why, you may ask, can we not fund *all* the ideas everyone has proposed for maintaining U.S. economic vitality in the face of rapidly increasing competition from other countries? Why can we not take advantage of *all* the research opportunities now available to us by virtue of new instrumentation, new computing power, and the mounting base of new information about everything from dark matter to social behavior? I believe we can do all the R&D we *need* to do, and very much of what we *want* to do, but I do not believe we can accomplish this the way we would *like* to do it, namely, by simply appropriating more federal funds.

Neither this administration nor any future one can escape the urgent demands of twenty-first-century realities. The struggle against terrorism is real and persistent. Climate change demands attention. Globalization is bringing the problems of countries around the world to our doorstep. And we have yet to address the looming crunch of entitlement programs in our own country—funded through the relentlessly expanding mandatory portion of the federal budget.

All these demands impact the domestic discretionary budget, which for decades has not grown as fast as the gross domestic product. It is an empirical fact that the science share of the discretionary budget has remained practically constant over time, so of course its share of GDP has fallen, too. Many science advocates, including probably most people in this audience, have used the resulting decline in ratio of federal research to GDP to argue for bigger federal science budgets. Because of the constraints on the discretionary budget, this argument will not be effective in the long run.

Two years ago in this forum, to repeat myself again, I argued that the ratio of federal science funding to GDP is not necessarily a meaningful indicator of a nation's science strength. I called for better benchmarks and a new "science of science policy" that would give us a surer foundation for setting priorities and better arguments for taking action. I am impressed and pleased with the response to that plea, not only by our own National Science Foundation, which has launched a program in the "social science of science policy," but also by the international community. The OECD—Organization for Economic Cooperation and Development—has acknowledged the need to have better data, better models, and better indicators that take into account the dynamic and global nature of research and development. Meanwhile, in the absence of a deeper understanding of cause and effect in the new era of globalized technical work, we need to be wary of reading too much into ratios and rates. Today, however, I want to make a different point.

Last October I gave a speech to the annual meeting of the Council on Governmental Relations (COGR) in which I expressed my concern about the mismatch between research capacity and the federal resources to sustain it. I claimed that "the universe of research universities has expanded to an economically significant size, by which I mean that the sum of financial decisions by its individual members has an impact on the resources available to any one of them. It is not quite a zero-sum game, but we have moved into a new operating regime where the limits of the 'market' for research university services are being tested." The doubling of the NIH budget that occurred, with everyone's blessing, over a five-year period ending in 2003, was an experiment in the rapid expansion of a broad but still well-defined scientific field. The most obvious lesson from this rapid growth is that it could not be sustained. There is a deeper lesson.

It is clear that the doubling has had a profound impact on the nation's biomedical research enterprise. It helps to think of this enterprise, and R&D activities generally, as a miniature economy with its own labor pool, markets, productive capacity, and business cycles. The response to the NIH doubling has been a step increase in research capacity, financed not only by the direct federal investment but by state governments and private sector sponsors eager to leverage this investment, not least to enhance competitiveness for additional federal funds. We now have an enlarged biomedical R&D labor pool—a new generation of researchers—who are populating new expanded research facilities and writing federal grant proposals in competition with the previous still-productive generation of their faculty advisors. And they are training yet another generation of new researchers who hope to follow the same pattern. I cannot see how such an expansion can be sustained by the same business model that led to its creation. The new researchers will either find new ways to fund their work, or they will leave the field and seek jobs in other sectors of the economy. This subeconomy is unregulated, and we can expect it to experience booms and busts typical of unregulated markets.

Under the stimulus of federal funding, research capacity as measured in terms of labor pool and facilities can easily expand much more rapidly than even the most optimistic projections of the growth rate of the federal research budget. New capacity can only be sustained by new revenue sources. In this connection, it is noteworthy that the federal research budget is dwarfed by private-sector research expenditures. Under the pressure of increased competition for federal funds, research universities are in fact forging new relationships with private sponsors, and I expect this trend to continue. The President's Council of Advisors on Science and Technology (PCAST) devoted a session in its recent meeting to reports by university, industry, and foundation leaders on modes of private-sector support for university-based research. Many universities are experimenting with new modes of interacting with industry and philanthropic organizations. Universities prefer sponsors who do not encumber their largesse with conditions, and the process of mutual accommodation with industrial sponsors may take time, but I believe accommodations are inevitable. The economics of university-based research are beginning to change to a new model with diversified sources of revenue.

Federal science policy should encourage this change. Not only will it enable an expanded research enterprise; it will also promote development of capacity in areas likely to produce economically relevant outcomes. Moreover, economists have documented a positive correlation between industrial research investment and national economic productivity, and to the extent this correlation indicates a causal relationship, increased industrial research will be good for the economy.

The message here is that federal funding for science will not grow fast enough in the foreseeable future to keep up with the geometrically expanding research capacity, and that state and private sector resources should be considered more systematically in formulating federal science policy. A possible precedent for federal action in this area may be found in the Bayh-Dole and Stevenson-Wydler legislation of more than twenty-five years ago. These acts gave ownership of intellectual property to the institutions in which it was developed with federal funds. Not only did it motivate federally funded research institutions to transfer technology to the private sector; it also created a dynamic that attracted private sector resources to the institutions. The level of industrially supported basic and applied research at universities remains low, however, relative to its potential.

Many precedents exist for private support of research, from the numerous societies formed to raise money for medical research on specific diseases to the remnants of the once robust system of industrial research laboratories. A remarkable example of private-sector involvement in institutions that also receive significant federal support is the string of more than a dozen research institutes endowed by the Kavli Foundation. One is located on the campus of SLAC—the Stanford Linear Accelerator Center—and forms a crucial part of the long-term strategic plan for that federal laboratory. Most other Kavli centers are on university campuses, and they represent a wide range of research topics, but always with a strong basic research orientation.

The links between federal, state, and private funding deserve more study. They are multiplying and growing stronger with relatively little federal encouragement. They appear to be building on foundations formed by federal funding, and there is no question that they could grow if encouraged by federal policies.

You will have your own ideas about how to fill the inevitable gap between the exponentially increasing research capacity and the much more slowly growing federal ability to satisfy it. Of all the policy issues to be discussed in today's forum, I think this one will be with us for the longest time and will have the greatest impact on how and what research is performed in our institutions. Research universities are responding with the creativity and entrepreneurial spirit characteristic of the U.S. economy as a whole. The annual AAAS forum is a good place to examine this movement, and the afternoon session on the "States' Expanding Role in Science and Engineering" is certainly relevant.

I am grateful to the American Association for the Advancement of Science for creating this forum, and inviting me once again to discuss these important issues. Thank you.

American Association for the Advancement of Science
33rd Forum on Science and Technology Policy
Keynote Address no. 7
Washington, DC
May 8, 2008

This is Marburger's final policy forum address. Leaving factual assessments of the Bush administration's legacy for science to other speeches—such as one written for *Physics World*[9]—he uses this one mainly to reflect on the role of a presidential science advisor. "An invitation to serve in a senior politically appointed position is not an invitation to bargain with the administration about significant policy issues," he said. "There isn't enough time for that. . . . Policies significant for science get shaped over a long period of time, very rarely overnight. But beginnings are important." He details the beginnings made during the Bush administration, as well as the dangers that he saw ahead. He encapsulated these in dramatic image, involving what he calls the "research commons." Marburger appropriated the phrase from the

---

[9]John H. Marburger III, "Has Bush Been Good for Science?" *Physics World*, November 2008.

biologist Garrett Hardin's paper "The Tragedy of the Commons."[10] The tragedy mentioned by the title is the tendency of an unregulated resource like ocean fish—something hitherto part of the global "commons"—to become depleted thanks to the (rational) actions of individual users. Invoking once again Sarewitz's point that R&D spending is a relatively constant portion of nondiscretionary spending, Marburger labels the R&D spending a "research commons." It is threatened today, he continues, by such things as earmarking, "which closely resembles the unregulated use of a common resource." While other countries exert stricter control over their research commons, the United States does not, and is unlikely to impose them. While our system is "ultimately more creative and flexible," Marburger warns that "it does permit digressions and distractions from policy goals." The beginnings initiated by the ACI and America COMPETES, he concludes, will need continued vigilance as the political actors change in the future.

—RPC

Thanks for inviting me to speak about the "Budgetary and Policy Context for R&D in Fiscal Year 2009," the theme of this morning's session. It's hard for me to believe this is the seventh time I've appeared at this policy forum as the president's science advisor and head of the White House science advisory apparatus. It must be just as hard for *you* to believe it, but here I am. Will it ever be possible for any of my successors to appear *eight* times? It would be difficult, given the lengthening time for clearances and the difficulty of finding qualified people at the beginning of each new administration to accept the heavy responsibilities, reduced compensation, and high workload—not to mention extraordinary public exposure—of senior federal appointments. "The Policy Context for R&D in fiscal Year 2009" begins immediately with the need for scientists and engineers who may be recruited by the next president to prepare themselves to say "yes" despite what may seem to be enormous downsides to this and other senior positions in the executive branch. Whoever becomes president, whatever party gains or loses power, and regardless of the specific policy environment in the next

---

[10]Garrett Hardin, "The Tragedy of the Commons," *Science* 162 (December 1968), 1243–1248.

administration, our government needs men and women who understand the science and engineering machinery in our society and are prepared to make it work for our nation.

An invitation to serve in a senior politically appointed position is not an invitation to bargain with the administration about significant policy issues, as many of my science colleagues tend to do. There isn't enough time for that. If you really want to help science, you help it regardless of the "policy context." Policies significant for science get shaped over a long period of time, very rarely overnight. But beginnings are important. Things need to get started immediately to take advantage of the sense of opportunity that always occurs with a new Congress and a new administration. I came to Washington nine months after the term had begun, and things were already rolling by that time. Of course, the main event that shaped my own first term in the White House was the terrorist attacks of 9/11. That was the theme of my first policy forum address seven years ago, and I don't intend to revisit it today. But I do want to make the point that there is important work to be done here, and much of it cannot be done through the advisory boards, commissions, committees and think tanks that academic scientists favor. Too many talented and experienced scientists and engineers who express interest in powering the nation's science machinery hang back from doing it as federal employees.

So let me praise the American Association for the Advancement of Science and a long list of partner organizations for sponsoring and expanding the all-important AAAS Fellows programs that not only increase Washington's technical sophistication but also spread appreciation among technical communities for how Washington works and the nature of its rewards for a scientist or engineer. My office, OSTP, has many former AAAS Fellows, and many of those spent some time as staff to members of the House Science Committee and other Capitol Hill committees relevant to our mission. OSTP itself has been a significant training ground for people who continue careers in government or policy positions in the private sector. I hope to see some AAAS fellowship alumni in important offices early in 2009.

The concept of a policy context is deeper than you might imagine. The concept of a policy itself is deeper than you might imagine. As I

PRESIDENTIAL SCIENCE ADVISOR II   195

mentioned in one of these talks years ago, my favorite science policy
paper is one that Daniel Sarewitz wrote in 2003 called "Does Science
Policy Exist?"[11] One of the positive events in science policy scholarship
during the past year is that Dan finally published his paper in the Summer 2007 issue of the National Academies journal *Issues in Science and
Technology*. The important insight of this paper is that, try as one might—
and generations of science advisors have labored on this—the process
of policy implementation in American government is shared over so
many uncoordinated actors that the policy behavior of the governmental science machinery is not only different from what policy makers
intended but sometimes difficult to trace at all. Anyone who wants to
evaluate the effectiveness of a policy first has to decide what the policy
is "on the ground" as opposed to some theory of it embedded in a law or
a strategic planning document. The actual outcome of the process of
policy proposal, authorization, and appropriation rarely resembles what
any particular stakeholder in the process hoped for at the beginning.

   That has certainly been true for the past two years during which two
different Congresses failed to appropriate science funding in patterns
that were worked out over months by their own authorization and appropriations committees. Most observers predict that once again the
forthcoming fiscal year will begin with a continuing resolution (which
translates as continuing irresolution). I think there is a danger of reading too much into this extraordinary failure of process. On Sarewitz's
model of real-world policy implementation, such things are to be expected in general.

   Another insight of Sarewitz's paper, to which I have repeatedly called
attention because of its importance for forecasting the outcomes of
policy struggles in Congress, or between the executive and legislative
branches, is the remarkable stability of federal R&D funding as a percentage of the domestic discretionary budget. Exactly how this happens
each year is somewhat mysterious. It's like trying to infer the laws of
thermodynamics from the behavior of the individual molecules in a
gas. The small-scale motion is chaotic and irreproducible, but the overall behavior always comes out the same.

---

[11]Sarewitz, "Does Science Policy Matter?"

The big picture of the national budget really matters. As Kei Koizumi [then-director of the AAAS R&D Budget and Policy Program] emphasizes at the beginning of his presentations each year, the discretionary budget is under constant and growing pressure from the large mandated portion of the budget. On top of this are huge fluctuations associated with events that may seem highly singular and catastrophic, such as war, hurricanes, and economic bubbles, but recur in nearly every administration. Regardless of the circumstances, the give-and-take of politics, including all the partisan dealing, all the lobbying, and all the local issues that intrude on the national scene, ends up giving research about the same fraction of the discretionary budget every year in administration after administration. The time series has bumps, but they rarely range outside a narrow band.

Many have complained about the impact of the 2008 omnibus bill on research. It did have negative effects, which I will mention in a moment, but the omnibus bill did provide increases for some important science areas, just not as much as the president's proposal, and much less than the appropriations bills being considered in the subcommittees. And of course, it did not reflect the priorities for funding either in the president's American Competitiveness Initiative, or in Congress's America COMPETES authorization bill. As the dust settles, however, once again research will have received approximately its usual slice of the pie.

I think this pattern will likely persist in future administrations. I think it will actually be difficult to match the increases in research funding that have occurred during the Bush years. There is much mythology about this, and much quibbling about definitions and what the numbers really mean, but overall there is a much greater amount of research money on the table today than there was at the beginning of the administration. Everyone has their own ideas about how it could have been distributed differently, both among fields of science and over the years. But there cannot be any question that this country has significantly boosted spending on research during this administration. The myths of downward trends in science spending are based on measures other than actual dollars spent. Patterns of U.S. science funding do show some disturbing trends, and they need to be fixed. But if we dwell only on those trends, I believe we indirectly raise false expectations that future administrations will be able to solve science funding problems simply by adding more funds to the pot. As always, there will be winners and losers.

Earlier this year I was asked by the *Harvard International Review* to write an article on "International Aspects of Climate Change." The editors explained that they were looking for a "theoretical" paper, which is easier for me to write than one requiring any real knowledge of the topic. In studying possible frameworks for thinking about how nations must behave to address the very serious and difficult challenge of global warming, I came across a 2002 National Research Council survey called *The Drama of the Commons*.[12] The reference here is to an influential paper by biologist Garrett Hardin in 1968 in *Science* magazine called "The Tragedy of the Commons." That term is now part of the policy vocabulary, and common resources scholarship has become an important area of social science research. The tragedy of the commons is the tendency of an unregulated resource (think of ocean fisheries) to become depleted as the result of rational choices made by individual users of the "commons." There are common resources, like fisheries, and common resource sinks for unwanted by-products, like the atmosphere. Hardin's paper is fascinating and should be mandatory reading by science policy students (it probably already is). The commons studies give us insights into what must be done to keep the commons productive.

As I was writing the paper on climate change, it occurred to me that Sarewitz's view of science funding has some of the characteristics of a resource commons. The funds are relatively stable and predictable from year to year, and they are exploited through the actions of multiple stakeholders, including executive agencies, members of Congress and their staffs, lobbyists, individual and organized science activists, think tank advocates, and many others who seize on science symbols to make their own case for a piece of the common resource pie. While there is at

---

[12]E. Ostrom, T. Dietz, N. Dolšak, P. C. Stern, S. Stovich, and E. U. Weber, eds., *The Drama of the Commons*, Committee on the Human Dimensions of Global Change, National Research Council Division of Behavioral and Social Sciences and Education (Washington, DC: National Academy Press, 2002).

least a management framework in the executive branch, of which my office and OMB are a part, that attempts to regulate the exploitation of this resource, there are no corresponding frameworks to limit the impact of the other stakeholders, including Congress. One result, of course, is the growth of earmarking, which closely resembles the unregulated use of a common resource.

This is the right place to acknowledge and praise the actions taken by the AAAS this year to treat earmarks in a systematic and straightforward way in their budget analysis. Perhaps Kei Koizumi, who speaks next, will say something about this. Two years ago I challenged OMB and AAAS to come together and work out a common approach to accounting for and report the effect of earmarks on the science budget. Last year OMB introduced an earmark website full of fascinating information. This year AAAS has made a useful study of the issue and has incorporated a good treatment in their reports. They estimate that congressionally designated, performer-specific R&D projects in 2008 total $4.5 billion and concluded that "in a tight budget environment earmarks once again crowd out hoped-for increases in competitively awarded research programs."

These earmarks are not simply whims of Congress. They come as a direct result of members of the science community, contractors, and others advocating for their projects. Many are worthy—let us stipulate that they are *all* worthy. But is this the most effective use of our resource commons? What guarantees the result is the best for the American people or is even, in some sense, sustainable? What guarantees that the most essential research ever gets done? Enlightened organizations representing large numbers of constituents here in Washington have made some efforts over the years to persuade their members that they need to advocate for broad programs, or for agency funding, and not for specific projects that seek to avoid review before receiving funds. They have not had much success. The jobs of many in this audience are part of a large and growing machinery aimed at bringing home the bacon for your organizations.

Earmarks are not the only aspect of the exploitation of the science funding commons. Various set-asides and incursions into the agency research missions have a similar effect. Like the SBIR [Small Business Innovation Research] tax on every agency's science budget that narrowly escaped a costly increase earlier this month, thanks to a floor action initiated by physicist-congressman Vernon Ehlers. Or the addition of new dimensions to program requirements related to education or outreach. The research commons is shrinking because of these incursions. Let's assume here, too, that they are all worthy. Most taxpayers look at these agency budgets—for NSF or the Department of Energy, for example—and assume the spending is for science. How is the science going to survive as the commons shrinks under these incursions?

I don't want to dwell at length on this problem, but it is very serious, and I think it helps to view it through the resource commons lens rather than as a problem of ethics. Curiously, Garrett Hardin's subtitle— sometimes presented as the single-sentence complete abstract—for his original and famous "Tragedy of the Commons" paper is "The population problem has no technical solution; it requires a fundamental extension of morality." So he, at least, thought the situation in extreme cases entailed moral considerations. Most social scientists who study these questions agree that commons must be managed under a consensual framework that regulates the behavior of the actors to avoid degradation of the resource.

During my tenure in Washington, I have seen three remarkable issues around which consensus among stakeholders has sprung up with an almost religious fervor: the successful doubling of the NIH budget between 1998 and 2003, the successful establishment of the new area of "homeland security research," and the as yet unsuccessful introduction of a set of actions, including selected research increases, to bolster future national economic competitiveness. The lack of success of the latter is quite unusual given the very wide and intense support for it across parties and sectors. I have said enough about the process failure that led to this result, but it is one of the most serious pieces of unfinished policy business begun in this administration.

There is no question that there is now a large and unhealthy imbalance among funds for various sectors of science, usually described as biomedical research versus physical science, engineering, math, and computer science. Federal support for the latter has languished for several decades. Now federal support for biomedical research has languished for half a decade. The budgets are still out of balance. We would certainly be better off today if the 109th Congress had passed its appropriations bills before going home. They would have begun to restore balance and start the building up of a technical workforce in fields badly needed for future progress in all parts of science and technology. This year, for the third year in a row, President Bush has proposed a path forward that aims to shrink the gap for critical fields where global competition is very strong.

Returning to the commons analogy, many other countries have forms of government that allow much stronger frameworks for managing the research commons. They have forms of central planning, regulation, and budget control that would never work in the United States. I think our system is ultimately more creative and flexible, but it does permit digressions and distractions from policy goals. Although corruption, waste, and ignorance divert resources in some other countries, some of them at least have systems that prevent substantial deterioration of the commons of public resources by the random incursions of individual well-meaning actors.

I think of the ACI and America COMPETES initiatives as unfinished business on an agenda that will continue to receive wide support because they address real and widely recognized problems. But maintaining that support will require continual effort by the organizations like the Council on Competitiveness and many others that have already worked so hard to make the case. The case has to be made over and over again as the political actors change.

There are other important challenges in our science coverage. Smaller agencies, like NIST, USGS, NOAA, and research units within other departments whose main mission is not science, often provide services that are critical to many other departments. Appropriators for these agencies have little incentive for heeding the needs of these other stakeholders. Think of the U.S. Geological Survey, for example, which Google and all the rest of us rely upon for land imaging. The increasingly urgent problem of water management depends on USGS research. Earth imaging programs of all sorts will need to grow in the future and be sustained through stable management structures within stewardship agencies. The visibility of NIST has been increased because of its inclusion in the ACI, but it has yet to receive the kind of boost its research budget requires to maintain its part of the innovation infrastructure. Within the huge Department of Defense, the basic research function has some of the characteristics of a small agency. DOD basic research funding has drifted since the end of the Cold War, much to the detriment of university-based engineering research. The good news for DOD research in the fiscal year 2009 budget proposal is a dramatic shift of resources toward the 6.1 basic research category. The reconstruction of the basic research function in DOD will take years and will only happen with constant support from the highest leadership levels. I give Secretary Gates [Robert Gates, secretary of defense] much credit for encouraging and supporting this initiative and hope his successors follow the same path. Finally, while NIH is certainly not small (although some of its institutes are), its budget cannot remain flat much longer. It will need to increase in a predictable, sustainable way in the future if we are to reap the value of our very substantial investment in biomedical research.

How are all these challenges to be overcome in a time of constrained budgets? In the immediate future, the best thing that could happen for U.S. science is for Congress actually to pass a budget for fiscal year 2009 as part of its regular business this year. I would like to see a bill that funds the president's request and finally launches a competitiveness initiative. The 2008 omnibus bill seriously wounded U.S. interests in high-energy physics and the international fusion energy project, ITER [International Thermonuclear Experimental Reactor], and it weakened our long-term prospects for competing successfully in a globalized technology intensive economy. That is the wrong signal to the American people, to the science and engineering communities, and to the world. The sooner Congress can pass bills moving us forward from this dreadful position, the better. As to the prospects for a supplemental budget that adjusts funding for fiscal year 2008, the issue is rapidly becoming moot because two-thirds of the year has passed already.

Timely passage of fiscal year 2009 budgets—which is, after all, what Congress is supposed to do—at this point would be the strongest bridge to the future.

I would like to thank AAAS for inviting me yet one more time to speak at this forum, and for sponsoring so many efforts to capture the strength of U.S. science and engineering for service to our country.

Epilogue
Marburger in the Bush Administration

All administration officials bear the burden of representation, for they are inescapably perceived as standing in for the institution. Their actions are thereby exposed to being cast as purely political actions, their faults and missteps ascribed to defective political ideology. Marburger's burden of representation was particularly tricky. He had to represent the scientific community and its findings to the administration, and he had to represent the Bush administration, which was free to accept or ignore these findings, to the scientific community. At the time, these two groups regarded each other with suspicion. As a result, Marburger was exposed to far more brutal and extensive criticism than other presidential science advisors.

Part of the reason for the harshness of the criticism was the historical moment. In 2000, George W. Bush won the electoral vote but lost the popular vote in a controversial process that was ended by the intervention of the U.S. Supreme Court. Marburger therefore became presidential science advisor at a time when many administration opponents were raising questions of the administration's legitimacy. Another reason is that these opponents were mounting a strong effort to depict the Bush administration as anti-science. They seemed to have plenty of ammunition. On two issues—global warming and stem cell research—the administration was not responding in the way that a strong consensus of the science community thought appropriate, and several other episodes involved scientific reports being changed for political reasons or individuals included or refused positions for their political views were also widely reported. Marburger, a respected scientist, loomed as a major obstacle to the credibility of this effort: if this Republican president was so antiscience, how was it possible that he had chosen a highly intelligent, scientifically respected, no-nonsense, articulate Democrat

PRESIDENTIAL SCIENCE ADVISOR II   203

as a science advisor? It is standard practice in Washington that, if something threatens your narrative, you seek to discredit it. Marburger was not
just difficult to discredit; he never shrank from pointing out the positive aspects of the Bush administration's impact on science. The following remark,
published in *Physics World* in November 2008, is typical:

> Today, the scientific enterprise in the U.S. is strong, highly productive and
> significantly greater than it was eight years ago. Contrary to popular my-
> thology, President Bush has devoted more attention to science and tech-
> nology in his official actions than most of his predecessors. Strains and
> imbalances exist among the various research fields, but the Bush admin-
> istration has initiated programs to address many of these on a prioritized
> basis. However, despite the magnitude of competing national needs and
> fiscal constraints affecting all domestic federal programs, science in the
> U.S. has moved forward substantially during the Bush years.[13]

Small wonder that Marburger became a lightning rod for attack, some of
it vitriolic and highly personal.

But another factor was Marburger's approach to his office. He went to
Washington as an advisor, not as an advocate. He took the lesson of Sarewitz's observations that the federal research budget has been constant—
and that federal science policy is essentially federal science budget policy—to
be that the most effective approach for a presidential science advisor and
OSTP director is not to seek more federal dollars for science or to champion
specific science agendas but to try to improve the way money flows through
federal channels. Marburger also drew a strong distinction between science
and politics. He thought that preserving the science-politics interface was
important, and he wanted to keep the OSTP on the science side of that
interface.

Shortly after he arrived in Washington, Marburger spelled out his reasons for this approach in a talk at the George C. Marshall Institute. He briefly
reviewed his previous jobs and remarked, "The curious thing about these
roles is that at no time did I regard myself as being a political actor. Yes, that
is naive, but it is true." He went on to argue why he thought this the right
approach to his current position:

---

[13]Marburger, "Has Bush Been Good for Science?"

Scientists do not regard themselves as political because they normally enter politics only in a peripheral way in pursuit of their own ends. . . . It is difficult for scientists to grasp the remarkably intuitive and indirect character of politics. Their—I should say "our"—instinct is to make things explicit, to expose cause and effect, and to work for clarity. We are result-oriented, our analyses are model-oriented, and we assume the models are stable. We believe in an objective reality governed by laws. Politics is none of these things. It seeks to conceal relationships, not for dubious reasons but as a matter of practical necessity. It is power-oriented and values personal loyalty above almost all else. It is opportunistic with respect to issues and methods, because it must be to get anything done. . . . These limitations of scientific culture with respect to politics lead to profound misunderstandings, grotesquely magnified by the media. The fact that scientists rely on politics to fund their programs but do not regard themselves as political and do not relate well to political behavior leads to mutual incomprehension across the science-politics divide. This is particularly true about the most political of all the transactions that occur across this divide, which is the getting of money for projects, but it also affects the next most important transaction, which is the giving of technical advice.[14]

Given these incompatibilities and misunderstandings, Marburger continued, it is no surprise that trouble often breaks out at the interface between science and politics. The trouble is compounded, he added, "when scientists are perceived by politicians to own a popular cause." All of this helps to explain why Marburger went on in the ensuing years to make sure his office respected the science-politics interface. As Richard Russell, the OSTP's deputy director for technology in the Bush administration, noted, "I think Marburger was the least political science advisor we've ever had" (personal communication).

Marburger understood that few science controversies are actually about science, and that appeals to science are often in the service of agendas where important ethical issues lurk in the background. "Even the sacred

---

[14]"Founders Breakfast Policy Briefing," presented at the George C. Marshall Institute, Washington, DC, June 23, 2003. John H. Marburger III Collection, Box 33, Stony Brook University Special Collections and University Archives, Stony Brook University, NY.

PRESIDENTIAL SCIENCE ADVISOR II

work of science can provide a vehicle for other issues, and this extra baggage is not always visible," he remarked in the Marshall Institute talk. Both the left and right can play this game—the right, for instance, when it invokes "women's health" or "a child's health" as a reason to seek to restrict abortion, the left when it invokes "patients' health" in seeking to lift restrictions on stem cell research. Marburger thought that he could best preserve and promote the authority of science in government if he kept his office out of such battles. This was the reason he had treated the Union of Concerned Scientists' "Restoring Scientific Integrity" statement as a set of factual claims rather than as a political document, and assiduously set out to determine the truth value of each, exposing which were false or misleading and setting out to fix the ones that were correct.

In September 2005, the *New York Times Magazine* published a profile of Marburger whose author interviewed critics of this approach. The author cited some science administrators as expressing grudging respect for Marburger in light of the inevitable compromises that one faces in such a position. The author also cited an unnamed "former Bush administration official" as making the (clearly hyperbolic) remark that the scientific community was lucky that Marburger had the position he did: "The choice is between Jack and a Neanderthal." But the bulk of the article gave voice to prominent scientists who viewed Marburger as collaborating with an administration that was guilty of "misuse" of science and of "systematically" manipulating scientific findings, and discussed the Union of Concerned Scientists "Restoring Scientific Integrity" document. "There were those who hoped that Marburger would tender his resignation in a show of solidarity," the article said. Marburger did not, and chided several of the individuals involved with acting in ways that hurt rather than helped the voice of science in administrative circles. When the author of the *New York Times Magazine* article asked Marburger about the actions of the scientific community, Marburger dismissed them as "shrapnel in the air," irrelevant to the task he was concerned with.

Science has a dual social role. It is a rigorous discipline with its own norms and vitality independent of the general culture—but it is also historically and practically important to the maintenance and ambitions of that general culture. This dual social role creates a tension in science administration between the desire to foster science apart from politics, and the desire to use it as an instrument for social ends. As Marburger frequently noted to his staff in assessing criticism from scientists on issues that had little bearing on their scientific fields, "Scientists have opinions like the rest of us." They are given increased standing in the media for their fact-based knowledge, but when they venture outside their chosen fields of science to opine on unrelated policy, their opinions are no different than those of other heartfelt supplicants who wish the country would function in a manner more to their liking.

The majority of the academic community, including the most politically vocal scientists, could not forgive President Bush his positions on abortion, gay marriage, social welfare spending, or the Iraq war. They therefore could never understand how one of their own could work for such an administration. Marburger, however, felt strongly that his role was to give the president the best scientific advice available, free from politics and personal opinion.

—RPC

6

The Science and Art of Science Policy

After stepping down as presidential science advisor in 2009, John Marburger published a book about quantum mechanics, *Constructing Reality: Quantum Theory and Particle Physics* (2011). He continued to promote the science of science policy, and he coedited *The Science of Science Policy: A Handbook* (2011). He wrote talks and essays about science policy and completed the first two chapters of this book. As we have seen, his perspective on science policy was that it involves far more than making decisions about the size and apportionment of science budgets. Indeed, little is to be gained by seeking larger injections of federal money into research, he felt, given the historical trend that this amount closely tracks the gross domestic product; however, there is much room for improvements in implementation, in the way that research money flows through the system. Marburger's chief aspiration as a science policy maker was therefore not to get more money into the system but to get the money we are currently putting into the system to do what we expect it to do better. In the previous chapters we have seen an array of strategies by which Marburger sought to do this: respecting the distinction between advocacy and advice, embedding science policy decisions in a recognizable narrative, heeding the Principle of Assurance, assuring the credibility of the policy maker, being unafraid to prioritize, and seeking to develop measures and other tools for making good decisions about policies. This chapter comprises a pair of articles that address what, at the end of his career, Marburger considered the two chief weaknesses of current science policy: a lack of tools and metrics to guide policy makers, and an unrealistic attitude among scientists and science policy makers that science is inherently authoritative in governmental circles.

Why Policy Implementation Needs a Science of Science Policy

How Marburger set about promoting the science of science policy provides a good illustration of how he thought science policy actors should act. Marburger had been germinating the idea for several years; already in his first keynote address to the American Association for the Advancement of Science (AAAS) policy forum, he mentioned the need "to develop and use the social sciences more effectively as a tool for public policy," as well as the need for "science-based science policy." The idea sprang from his conviction that his job was to advise the president about science issues. Much of this advice was in response to specific events or controversies, such as tsunamis or nuclear power, but Marburger's vision was farther reaching. A presidential science advisor's charge, he felt, included making sure the portfolio of science funding is properly managed and in the country's best interest. If, as Marburger liked to quote Daniel Sarewitz, "federal science policy is largely played out as federal science budget policy,"[1] a corollary is that making sure the budget flow achieves what is intended is science policy of the highest sort.

But how could a policy actor acquire that assurance? Marburger's sense that science policy was being driven by anecdote and advocacy rather than by the kinds of models that economists use troubled him increasingly throughout his service during the Bush administration's first four years. After the 2004 election, he began to address the issue, collecting and discussing, with Office of Science and Technology Policy (OSTP) deputy director for science Kathie Olsen and National Science and Technology Council executive secretary Ann Carlson, available resource literature on econometrics and on the impact of science policy. These discussions crystallized his call, in his AAAS policy forum keynote address of April 21, 2005, for a "social science of science policy," further publicized in his May 2005 editorial in *Science* magazine.

That was just the beginning. Marburger used personal leverage and professional contacts to gather more information and promote the idea in steadily widening circles. He scheduled a meeting with Ben Bernanke, the newly appointed chairman of the Council of Economic Advisers and soon to

---

[1]Daniel Sarewitz, "Does Science Policy Matter?" *Issues in Science and Technology* (Summer 2007), http://issues.org/23-4/sarewitz/.

THE SCIENCE AND ART OF SCIENCE POLICY   209

become chairman of the U.S. Federal Reserve, to pick Bernanke's brain
about the genesis and use of econometric models and the potential for developing such models to evaluate investments in basic research. Bernanke
was not terribly impressed, saying something to the effect that "I don't think
that the government should be funding anything that does not have demon-
strable economic consequences." Still, Bernanke provided Marburger with
some information about resources on econometrics.

Marburger also sought international perspectives. A month after his
2005 AAAS keynote address, he broached the issue of developing a science
of science policy at a meeting of science ministers and senior government
officials concerned with science from the G-8 nations.[2] These meetings are
sponsored by the Carnegie Group and are typically held once or twice a year;
this one was on May 27, 2005. He, Olsen, and Carlson then formulated a
proposal for the Organization for Economic Cooperation and Development
(OECD) to host a workshop on science indicators at its meeting in Helsinki
the following year.[3] The proposal was approved at OECD's July 2005 Global
Science Forum meeting in Copenhagen, and the workshop was held in July
2006 at the forum's next meeting in Helsinki.[4]

Meanwhile, Marburger pushed for more development of science indicators and science policy research in the United States. He approached the
National Science and Technology Council (NSTC)—the body that coordi-
nates science and technology policy across the various groups that make up
the federal research and development enterprise—and arranged for them
to organize an interagency working group on "science indicators." These
working groups typically meet once or twice to formulate a recommendation, though for bigger issues—as in this case—the recommendation is often to convene a more permanent and formal interagency task group to

---

[2]John Marburger, "Toward an Improved Framework for Understanding Science In-
dicators," paper presented at the G-8 meeting, May 27, 2005. John H. Marburger III
Collection, Box 35, Stony Brook University Special Collections and University Archives,
Stony Brook University, NY.

[3]"Proposal for a Workshop on Fundamental Science Indicators and International
Benchmarking," submitted to the thirteenth meeting of the OECD Global Science Fo-
rum by the U.S. delegation. John H. Marburger III Collection, Box 35, Stony Brook
University Special Collections and University Archives, Stony Brook University, NY.

[4]The workshop was titled "GSF Workshop on Science of Science Policy: Developing
Our Understanding of Public Investments in Science," held July 12, 2006, in Helsinki,
Finland.

craft a long-term recommendation. In June 2005, Marburger convened a meeting at the OSTP with representatives of the National Science Foundation (NSF) to help set up such a task group, which met for the first time in August 2006. Two years later the task group issued its report, containing a roadmap and numerous recommendations.[5]

Marburger also pushed the NSF to promote research into improving policy indicators and generating additional ones and encouraged them to create a new program devoted to the subject. For this to happen, he had to involve the all-important Office of Management and Budget (OMB)—and here his careful cultivation of the OMB came into play. In the fall of 2005, he made sure that the joint OMB-OSTP budget letter—at that time still affectionately known by the local science policy community as the Jack Daniels memo (because it used to be signed jointly by Jack Marburger and Mitch Daniels), included language calling for agencies to consider developing programming to investigate ways to improve science indicators and their use in formulating science policy. Encouraged by this sign, the NSF submitted a proposal for a "science of science and innovation policy" program in its 2007 budget request. The request was approved, and the agency was able to begin funding programs in the science of science policy in 2007.

Marburger's actions were not without critics and skeptics. Some thought that it was hopeless to seek metrics of the sort Marburger desired because the political ecosystem was so complex and ever-changing that no one variable or set of variables could give you meaningful insight. Others thought that the current social science activity was already addressing the issue. For instance, Susan Cozzens, professor of public policy and director of the Technology Policy and Assessment Center at Georgia Tech, criticized Marburger's description in his 2005 AAAS keynote policy forum talk of the "social science of science policy" as being a "nascent" field. "For decades," she wrote, "U.S. social scientists, working with scientists and engineers, had been producing policy-relevant methods, models, and results to help characterize the science and engineering enterprise."[6] This is partly true yet coming from an academic perspective misses the point of Marburger's actions.

---

[5]National Science and Technology Council, "The Science of Science Policy: A Federal Research Roadmap" (November 2008), http://cssip.org/docs/meeting/science_of _science_policy_roadmap_2008.pdf.

[6]Susan E. Cozzens, "Science and Innovation Policy Studies in the United States: Past and Present," Working Paper 53 (Georgia Tech Ivan Allen College School of Public Policy, March 2010), http://hdl.handle.net/1853/32981.

THE SCIENCE AND ART OF SCIENCE POLICY   211

Marburger was indeed aware of the social science literature on science policy from his earliest discussions with Olsen. What troubled him was that
this literature was having little or no impact. His efforts were geared toward
making sure that social science research had a high enough profile that it
would be incorporated into the judgments of policy actors. Having worked
at a university and a laboratory, he was painfully aware that scholarship
does not naturally get picked up by government decision makers, even
those to whom it was relevant (a point that he explored further in the article
on authority included below). Social science research on science policy,
Marburger felt, needed a push, it needed to become a "movement." For this
to happen, the field needed much more substantial funding, a critical mass
of ongoing activity, and a name.

" 'Policy' is no more than a set of guidelines to shape actions toward
ends," Marburger told a Brookings Institute conference on policy in 2007.[7]
"Having a policy implies that you have a goal and an idea how to achieve it."
He continued:

> To a scientist that sounds like having an hypothesis about the system of
> people and resources engaged in the activities subject to the policy. And
> the power of science comes from the imperative to test hypotheses
> against reality. I view policies as hypotheses, embodying theories of eco-
> nomic or social behavior that may or may not be correct. If policies are
> more than empty generalizations, they have consequences, and those
> consequences must be tested against our expectations if they are not to
> be counterproductive. Policy formation must have an empirical basis,
> and policy execution must include the monitoring and analysis of conse-
> quences. These are features of what I have called the science of science
> policy, but they apply to all policy.

After a remark about how difficult it is to relate investments in science to
measurable beneficial outcomes, Marburger said:

> Most advocacy cites data in support of its causes, and the increasing
> power of information technology has made a flood of data available for this
> purpose. Policy makers cannot simply accept the assertions of advocates.

---

[7]"Four Themes in Science Policy," paper presented at the Brookings Institute Policy
Conference, Washington, DC, June 11, 2007. John H. Marburger III Collection, Box 36,
Stony Brook University Special Collections and University Archives, Stony Brook Uni-
versity, NY.

They need a disciplined approach to data and to the development of models that give meaning to the numbers. This is especially true in our era when basic patterns of economic activity are changing rapidly, including the part of the economic universe that is science. Intuitions based on past policy successes are not reliable today if only because the scale, pace, and geographical distribution of nearly all productive activities have changed. I know from my own direct experience that the practice of research today is vastly different from the past. Information technology and globalization have changed everything. It is more important than ever to relate policies to the actual working of the organizations and individuals whose behavior we are attempting to influence.

The following selection is Marburger's contribution to *The Science of Science Policy: A Handbook*, slightly edited.[8] In it, he points toward the kinds of metrics that will have to be developed to put science policy on a firmer foundation, and to the kind of scholarly work that will have to be done to create these metrics. Such metrics will be difficult to develop, but for him difficult does not mean impossible.

—RPC

1. Introduction

My perspective on the needs of science policymakers was strongly influenced by my experience as science advisor to the president and director of the Office of Science and Technology Policy (OSTP) during the two terms of President George W. Bush's administration (2001–2009). Watching policy evolve during an era of dramatic economic change, deep political divisions, and high visibility science issues made me aware of weaknesses in the process that will not be remedied by simple measures. Science policy studies illuminate the problems and undoubtedly propose options for addressing them but will not by themselves solve them. The growth of science policy studies as an academic discipline nevertheless

---

[8]John H. Marburger III, "Why Policy Implementation Needs a Science of Science Policy," in *The Science of Science Policy: A Handbook*, ed. Kaye Husbands Fealing, Julia I. Lane, John H. Marburger III, and Stephanie S. Shipp (Stanford, CA: Stanford University Press, 2011), 9–22.

THE SCIENCE AND ART OF SCIENCE POLICY   213

provides intriguing opportunities for steady improvement in the management of the national science and technology enterprise.

2. Science Policy Challenges in the Executive Branch

The structure and responsibilities of the OSTP have changed very little since the office was established by Congress in 1976, and its basic form had been in place since President Eisenhower appointed the first full-time science advisor in 1957. The advisors have always played two roles: (1) advising the president and his senior policy officials on all technical matters that reach the executive level, and (2) coordinating, prioritizing, and evaluating science and technology programs throughout the executive branch of government. A third responsibility is obvious but rarely discussed: the responsibility, shared with many others, of seeing policies through to their successful implementation. Each of these roles and responsibilities has its challenges, but none is more difficult than the third. The nature of this difficulty severely constrains strategies to overcome it, but the "science of science policy" movement is a potentially powerful tool for policy implementation. To appreciate why requires some background on the federal decision-making machinery. Science policy has many dimensions, but here I will focus specifically on the implementation of initiatives that require authorization and appropriations by Congress.

Policies are guides to action. Therefore strategies for implementation are nearly always embedded in policy proposals, often implicitly. The actions required to transform a policy idea into a desired result occur in stages. The early stages are successive expansions of the group of agents and stakeholders whose endorsement is needed to launch the initiative. Later stages focus on the management of the program, feedback of information about its success or failure to the policy level, and subsequent policy level actions responsive to the feedback. Together these stages comprise the natural cycle of planning, implementation, evaluation, and improvement that applies to all systematic efforts to accomplish defined objectives. Science projects normally occur within a larger framework administered by an organization or a governmental agency. My interest here is in frameworks for science and technology programs at the highest policy level and early-stage actions required to launch them.

The complexity of the U.S. federal science establishment is notorious. The executive branch carries out the business of government through a large number of departments and agencies, many of which today have a research arm and an employee in the role of chief scientist. Nineteen of these organizations are designated by an executive order as members of the interagency National Science and Technology Council (NSTC), managed by the OSTP. Twenty-five of them participate in the interagency National Nanotechnology Initiative (NNI), and thirteen each participate in the Global Change Research Program (GCRP) and the Networking and Information Technology Research and Development (NITRD) program.[9] Among the fifteen departments and fifty-six "Independent Establishments and Government Corporations" listed in the current edition of the *U.S. Government Manual*, only one, the National Science Foundation (NSF), is fully devoted to the conduct of science, including research fellowships and science education.[10] All of the other science organizations are therefore embedded within larger departments in which they compete with other functions for money, space, personnel, and the attention of their department secretary or administrator. Two of the largest science agencies, NASA (National Aeronautics and Space Administration) and NSF, do not report to a cabinet-level administrator and rely on the OSTP to make their case in White House policy-making processes.

The dispersion of research through such a large number of agencies was a weakness already recognized in the 1940s by Vannevar Bush, who urged consolidation into a single basic research agency.[11] That ef-

---

[9]For a description of these programs, see the OSTP's website,www.whitehouse.gov /administration/eop/ostp.

[10]The *U.S. Government Manual* is available at www.gpoaccess.gov/gmanual/.

[11]*Science: The Endless Frontier; A Report to the President by Vannevar Bush, Director of the Office of Scientific Research and Development, July 1945* (Washington, DC: U.S. Government Printing Office, 1945), chap. 6, www.nsf.gov/about/history/vbush1945.htm.

THE SCIENCE AND ART OF SCIENCE POLICY   215

fort led ultimately to the creation of the National Science Foundation in 1950, but the consolidation included only a small fraction of the then-existing federal research portfolio. The bureaus that became the National Institutes of Health (NIH), NASA, the Department of Energy (DOE), and the Department of Defense (DOD) research entities remained separate and today have research budgets comparable to or greater than the NSF. Many smaller science agencies within cabinet departments, such as Commerce, Agriculture, and Interior, also remained separate. The challenges of managing multiple science enterprises in the executive branch motivated the development of bureaucratic machinery in the 1950s to avoid duplication, fill gaps, and preserve capabilities serving multiple agencies.[12] The White House Office of Management and Budget (OMB) has the greatest authority in this role, and the OSTP works with the OMB to establish priorities and develop programs and budgets for the science and technology portions of all the agencies.

The OMB itself is divided into five relatively independent divisions (four during my service), each of which manages a significant portion of the overall science and technology activity.[13] This creates a challenge within the executive branch for initiatives that cut across the major science agencies. The NIH, NSF, DOE, and DOD research budgets are each developed in separate OMB divisions. Budget officials work hard to protect their independence, and they attempt to insulate their decisions from other White House policy offices. Major policy decisions are made through a deliberative process among White House and cabinet officials—always including the OMB—that narrows issues and choices for ultimate action by the president. Only a few issues, however, can receive such high-level attention, and most decisions about science and

---

[12]For the early history of White House coordination of science, see "Impacts of the Early Cold War on the Formulation of U.S. Science Policy: William T. Golden," W. Blanpied, ed. (Washington, DC: AAAS, 1995), http://www.aaas.org/page/impacts-early-cold -war-formulation-us-science-policy-william-t-golden.

[13]For OMB organization, see www.whitehouse.gov/omb/assets/about_omb/omb _org_chart.pdf.

technology policy are negotiated within agencies and among the various White House policy offices.

This executive branch machinery is complex but reasonably well defined and understood by the bureaucracy. However, it is not always well understood by the political appointees in each agency whose tenure is often less than a four-year presidential term. Their involvement in the process adds to the element of randomness always present in agency responsiveness to presidential direction, but the political appointees also reduce the impedance mismatch between the volatile political leadership and the cultural inertia of the bureaucracy. Designing high-level policies within the executive branch so they will actually be implemented requires detailed knowledge of the political and bureaucratic cultures of the agencies that will be responsible for implementation. Because these cultures depend on personal qualities of the agency leadership and specific historical tracks to the present state of the agencies, policy analysis and design are not well defined. For this and other reasons common to complex organizations, the behavior of the agencies is not sufficiently predictable to guarantee that a policy, once launched by presidential directive or executive order, will follow an anticipated trajectory. The usual government remedy for the consequences of this uncertainty is to establish stronger coordinating organizations at the top, such as national coordinating offices (e.g., for the NNI and the NITRD), "czars," or presidential commissions. The OSTP itself has become increasingly effective in this role through refinement of the NSTC structure over several administrations. Notwithstanding these arrangements, the agency line management is on the job continually and has more resources than the relatively small executive office of the president to influence the action environment. I had this phenomenon in mind when I described the origins of the Bush administration's vision for space exploration to the 2009 "Augustine Committee" that reviewed NASA's human space flight plans: "The final [space] policy document was a compromise between contrasting policy perspectives offered by NASA and by the White House policy advisors. In subsequent presentations to Congress and to the public, NASA representatives emphasized the NASA view of the *Vision*, which began to appear even during the

THE SCIENCE AND ART OF SCIENCE POLICY   217

policy formation process through leaks to the media serving the space community."[14]

3. Legislative Impact on Science Policy Implementation

The legislative branch has no executive machinery to resolve the random forces that influence its own operations. It responds to the president's budget proposals with two dozen very independent appropriations subcommittees, as well as a large number of authorizing committees and subcommittees. No organization, such as the OMB or the OSTP, monitors or attempts to enforce policy consistency across the hundreds of bills passed in each Congress, much less the thousands of bills that are introduced and debated. Offices such as the Congressional Budget Office (CBO), the General Accountability Office (GAO), and the Congressional Research Service (CRS) are informational only and have no authority over the 535 members of Congress. These offices are influential, however, through the quality and perceived objectivity of the information and analyses they produce. The CBO analyses are particularly successful in fostering a consensus on the financial aspects of legislative proposals.

Legislative funding for science and technology originates in nine of the twelve different appropriations subcommittees in each chamber, for none of which is science the sole or even majority category of funding. The "big five" science agencies—NIH, NSF, DOE, NASA, and DOD—are funded by four different appropriations subcommittees. Each subcommittee has its own staff, whose members' voices are more influential than many executive branch policymakers in establishing priorities and programs among the executive agencies. And each subcommittee is a target for its own army of advocates, lobbyists, and activist individuals whose influence is difficult to trace but highly significant.[15]

---

[14]These remarks, as well as other supporting documentation, may be found on the website for the "Review of U.S. Human Space Flight Plans Committee," www.nasa.gov /offices/hsf/meetings/08_05_meeting.html.

[15]One measure of external influence is the number of earmarks on the appropriations bills. The OMB counted 11,524 earmarks on the 2008 budget bills. See http:// earmarks.omb.gov/earmarks-public/.

Sections of bills are often drafted by lobbyists or constituents of a sub-committee member or the chairperson. The subcommittees are sub-stantially "stovepiped," with little incentive to coordinate action except on highly visible multiagency issues such as climate change or energy policy. The authorization and appropriations bills give surprisingly spe-cific and sometimes conflicting direction to agencies, substantially and routinely invading the president's constitutional prerogative to manage the executive branch.

This complex and unpredictable field of action leads to inefficiencies and perverse distributions of resources that create a continual irritant, if not a threat, to America's otherwise very strong research and develop-ment (R&D) enterprise. One extreme example is the tortuous history of the initiative to enhance U.S. economic competitiveness in the second Bush administration. In 2005 a wide consensus developed within the U.S. science and technology community that U.S. economic competi-tiveness was threatened by neglect of the nation's "innovation ecol-ogy."[16] The president responded with the "American Competitiveness Initiative" (ACI) in his 2006 budget proposal to Congress. The 110th Congress authorized its own response (largely consistent with but more generous than the ACI) in the "America COMPETES Act of 2007 (ACA)."[17] Congress, to the great surprise and consternation of the com-munity, failed to fund the program because of a stalemate with the president regarding his insistence that the total budget (not just for R&D) not exceed his top line. For three years the initiative languished until the Bush administration expired and the 111th Congress substan-tially funded the initiative, and much more, along with the American Recovery and Reinvestment Act of 2009. In subsequent budget sub-missions, the Obama administration has generally supported the main provisions of the ACI and the ACA. During the final Bush administra-tion years, science budgets reflected the priorities of the appropriations

---

[16]See National Academy of Sciences, National Academy of Engineering, and Insti-tute of Medicine, *Rising Above the Gathering Storm: Energizing and Employing America for a Brighter Economic Future* (Washington, DC: National Academies Press, 2007), www.nap.edu/catalog.php?record_id=11463.

[17]Some information about the ACA can be found at the House Committee on Sci-ence, Space, and Technology website, science.house.gov/.

THE SCIENCE AND ART OF SCIENCE POLICY   219

committees, not the executive branch and its scientific advisory panels. Politics played a dominant role in this saga, but other factors also were significant, including the fact that the ACI and, to a lesser extent, the ACA identified explicit priorities. Major science agencies such as NASA and the NIH were not included in the initiative. Consequently, Congress was somewhat insulated from criticism for its failure to act because important science constituencies excluded from the initiative remained silent. During this period Congress continued to add large earmarked amounts to the R&D budgets, but few were in the prioritized programs.

In the longer run the "competitiveness campaign" resulted in important changes in the pattern of appropriations for science, as did the earlier campaign to double the NIH budget in the 1990s. In both cases the late stages played out in a new administration affiliated with a different political party, which suggests that a sufficiently broad, bipartisan campaign can succeed regardless of which party is in power. Such broad consensus is difficult to achieve, which is why it usually occurs only in the face of a national crisis: notable federal R&D funding spikes occurred during World War II and after the 1957 Soviet *Sputnik* launch, while others followed the oil embargo in the 1970s and perceived Cold War urgencies (e.g., President Reagan's Strategic Defense Initiative) in the 1980s. The NIH and competitiveness initiatives were not propelled by similarly dramatic events, but champions for both campaigns developed cases based on "disturbing trends," including lagging rates of R&D investments compared to other countries, discouragement or lack of preparation of potential young scientists, and shortsighted abandonment of basic research in favor of applied research and development.[18] These and similar arguments were taken up by advocacy organizations of which some, such as Research America and the Task Force on the Future of American Innovation,[19] were formed for the purpose. Prominent figures were recruited, op-eds were written, conferences and

---

[18]See n. 9 and J. M. Bishop, M. Kirschner, and H. Varmus, "Policy Forum: Science and the New Administration," *Science* 22 (1993), 444–445.

[19]For Research America, see www.researchamerica.org; for the Task Force on the Future of American Innovation, see www.innovationtaskforce.org/.

"summits" were held, and ultimately, government responded. In the absence of a precipitating event, the advocacy communities worked to create a sense of national crisis to motivate the process.

4. Toward a Firmer Foundation for Science Policy

I too participated in the competitiveness campaign in my official capacity, encouraging an early report by the President's Council of Advisors for Science and Technology in 2002, overseeing the OSTP's role in crafting the ACI, and giving many supporting speeches. My direct experience with basic and applied research programs in diverse fields over four decades convinced me of the importance of these goals, but I was uneasy regarding the case put forward by the advocacy community. I thought the "disturbing trends" needed attention, but I was not convinced either that they would lead to the feared consequences or that the proposed remedies would work as advertised. On some issues, such as the status and future of the scientific workforce, there were deep uncertainties ("Do we have too many scientists/engineers in field X, or too few?").[20] My policy speeches from 2005 and thereafter expressed my frustration over the inadequacy of data and analytical tools commensurate with science policymaking in a rapidly changing environment.[21]

Given the complex and unpredictable systems of executive branch agencies and congressional subcommittees, no deployment of czars, commissions, or congressional offices will guarantee that rational and coherent policy proposals will be implemented. Congress departs reliably from the status quo only in response to widely perceived national crises, or when impressed with a broad consensus among multiple constituencies. If the consensus is created by advocacy alone, then there is no assurance that the proposed solution will achieve the desired end, even if the problems it addresses are real. Moreover, advocacy-based

---

[20]D. Kennedy, J. Austin, K. Urquhart, and C. Taylor, "Supply without Demand," *Science* 303 (2004), 1105.

[21]See J. H. Marburger, "Wanted: Better Benchmarks" [editorial], *Science* 308 (2005), 1087.

THE SCIENCE AND ART OF SCIENCE POLICY   221

consensus has never reached across all of the fields of technical endeavor that draw funding from the overall R&D pot. Attempts to prioritize among fields or agencies are extremely rare and never well received
by the scientific community. Past campaigns focused on selected fields
that were perceived to be relevant to the crisis at hand and ignored the
others.

Can a sustained science policy consensus develop that is strong
enough to influence the government machinery and wide enough to
encompass all of the disparate but linked technical endeavors that federal funds support? There is some hope. The National Academy of
Sciences (NAS) offers high-quality advice in every relevant technical
field, and the products of the National Research Council (NRC) carry
substantial weight with all of the actors in the complex process described earlier. The NRC reports avoid blind advocacy but are nevertheless assembled by teams of scientists and engineers, nearly always in
response to a specific narrow charter negotiated by the agency request-
ing, and funding, the study. Even a report as focused as *Gathering
Storm* avoided specificity regarding the relative importance of different
fields of science. When NAS president Frank Press urged his colleagues
in 1988 not to leave key priority decisions to Congress, he was roundly
criticized by his own community.[22] The only source of high-level policy
analysis that is relatively free of the biases of advocacy and self-interest
is the community of social scientists and others who analyze science
and technology policy as an academic field of study. And it is in the
growth of this community and its products that the greatest hope lies
for developing rational and objective policy perspectives that all parties
to the national process can draw upon. The collective endeavor of this
community is what I understand to be the science of science policy.

I became acutely aware of the inadequacy of available science policy tools following the terrorist attacks of September 11, 2001. These

---

[22]For a useful review of priority setting in science, including remarks on Press's
1988 annual president's address, see chap. 5 of U.S. Congress, Office of Technology Assessment, *Federally Funded Research: Decisions for a Decade*, OTA-SET-490 (Washington, DC: U.S. Government Printing Office, May 1991).

actions sparked a strong patriotic response in the science and engineering communities. Along with the desire to respond aggressively to terrorism came a wave of uneasiness about the impact on science of demands for increased homeland security. These included an immediate tightening of visas for students and visiting scientists, regulations on handling "select agents," or substances of likely interest to terrorists, concern over the release even of the nonclassified research results that might assist terrorism, and the possible diversion of funds from existing science programs to new efforts related to homeland security. All of this was new to the nation's technical communities, and sorting out the issues and options consumed huge amounts of time in studies and meetings. Among many other policy questions I was asked at the time was one raised by the National Science Board (NSB).

The NSB's Task Force on National Workforce Policies for Science and Engineering invited me to address its June 2002 meeting on the topic "Impact of Security Policies on the Science and Engineering Workforce." This was a reasonable request given the nature of the new policies, but it created a dilemma for me. Although I had no information on which to estimate the impact, the prevailing wisdom in the academic community was that it would be negative. I could think of reasons to reinforce that conclusion, but I was also aware of the complexity of the technical workforce picture that was changing rapidly because of globalization and profound development in China and India. Should I speculate? My extensive experience in a research university and national laboratory administration gave me the confidence to offer an opinion. Or should I point to the much larger issue of our ignorance about such impacts and what we would need to remove it? To quote directly from my notes for that meeting:

> The fact is, I do not know what the impact of security policies will be on the science and engineering workforce. Part of the reason for this—the least important part—is that the security policies are in a state of flux. Another part is that the impact will be psychological as well as instrumental, and psychology is not part of our predictive model. The most important factor, however, is that there is no reli-

THE SCIENCE AND ART OF SCIENCE POLICY   223

able predictive model for the workforce response to any particular
driving force, such as a change in policy affecting student visas.

If there are such models, they seem to be implicit in the types of
data we collect and the manner we choose to portray them. When I
see graphs and tables relating to workforce, I have the impression
they are answers to questions whose significance is either so well
known to experts that no further discussion is required, or so completely buried in history that no further discussion is possible. I
understand the need to collect the same data year after year so comparisons can be made and changes depicted accurately in the course
of time. But I am not at all confident that the right questions are being asked or answered to provide guidance for action. We have workforce data that I do not understand how to use, and we have workforce
questions whose answers would seem to require more than merely
data.

My idea at the time was that the NSB, which oversees the production of
the important Science and Engineering Indicators reports,[23] should
consider the task of building a new workforce model that might make it
possible to answer questions such as the one they asked me: "What do
we expect from a technical workforce model?" My responses to the
board follow:

I know what I expect from a model. I expect it to give policy guidance. I want to be able to assess the impact of a change of policy on
the technical workforce. . . . What is the impact of a student loan for-
giveness program? Of a scholarship program? Of a change in the
compensation structure of researchers, faculty members, technical
staff? Of an increase in sponsored research funds in some field? Of a
change in graduation rates in certain fields among certain sociological groups? Ask all these questions with respect to the nation in
which the changes are postulated to occur. It must be a global model,
because the workforce we are speaking of has global mobility. It

---

[23]See National Science Foundation, "Science and Engineering Indicators," www.nsf
.gov/statistics/seind/.

must take into account the effect of incentives, and the correlation of this effect with sociological parameters.

Above all, the model cannot be simply an extrapolation based on historical time-series data. The technical workforce is responding to factors that are changing too rapidly to be captured by historical data. And yet the model does not have to predict everything with perfect accuracy. What we need is the ability to estimate specific effects from specific causes under reasonable assumptions about the future. . . . Does it make sense for us to launch a project to model the global workforce with the aim of producing policy guidance? We need an action-oriented *workforce project* that seeks to define the technical workforce problem in a broad way, and to exploit the power of modern information technology to produce tools for policy guidance.[24]

I knew at the time that this was not a task the NSB was prepared to undertake, but I wanted to signal my concern about an issue that seemed to threaten the credibility of all policy advice. In the face of grave national challenges we were relying on anecdotes and intuitions and data disconnected from all but the most primitive interpretive frameworks. While I was prepared to accept responsibility for my policy recommendations, the scientist in me recoiled from the methodological weakness of this approach. If we think empirically based research is essential for learning about nature, or making useful products, then why should we not encourage research to build empirically validated foundations for effective science policy?

5. A New Mandate for Science Policy Studies

By 2005 I had concluded that no single project or study would address the need for better science policy tools. My keynote speech to that year's American Association for the Advancement of Science (AAAS) science policy forum compared our situation to that of economic policymakers who

---

[24]Prepared remarks for the meeting of National Science Board Task Force on National Workforce Policies for Science and Engineering, June 28, 2002.

THE SCIENCE AND ART OF SCIENCE POLICY   225

have available . . . a rich variety of econometric models, and a base of academic research. Much of the available literature on science policy is being produced piecemeal by scientists who are experts in their fields, but not necessarily in the methods and literature of the relevant social science disciplines needed to define appropriate data elements and create econometric models that can be useful to policy experts. . . . These are not items that you can just go out and buy because research is necessary even to frame an approach. This is a task for a new interdisciplinary field of quantitative science policy studies.[25]

The following year I addressed the "Blue Sky II" conference of the Organization for Economic Cooperation and Development (OECD) on "What Indicators for Science, Technology and Innovation Policies in the 21st Century?" Once again I drew a comparison with economic models:

Unfortunately, in our era of dynamic change, the empirical correlations that inform the excellent OECD analyses of economic performance are not very useful to science policymakers as guides to the future. They are not model in the sense that they capture the microeconomic behaviors that lead to the trends and correlations we can discover in empirical data. Take, for example, the production of technically trained personnel in China. China is producing scientists, mathematicians, and engineers at a prodigious rate. As a scientist and an educator, I tend to approve of such intellectual proliferation. As a policy advisor, I have many questions about it. How long, for example, can we expect this growth rate to be sustained? Where will this burgeoning technical workforce find jobs? What will its effect be on the global technical workforce market? Is it launching a massive cycle of boom and bust in the global technology workforce? Historical trends and correlations do not help here. Nor, I am afraid, does simply asking the Chinese policymakers what they intend. They also

---

[25]See the 2005 Keynote Address to the AAAS 30th Forum on Science and Technology Policy, chapter 5 of this volume.

need better tools to manage the extraordinary energy of their society. We need models—economists would call them microeconomic models—that simulate social behaviors and that feed into macroeconomic models that we can exercise to make intelligent guesses at what we might expect the future to bring and how we should prepare for it.[26]

By the time of Blue Sky II, the National Science Foundation had already held workshops and issued a prospectus announcing the formation of a new program in the social science and innovation policy (SciSIP) and solicited proposals the following year.[27] The president's FY2007 budget proposal to Congress included funds to launch the program, and several awards were distributed beginning in calendar year 2006. The current (2010) synopsis of the program indicates support for "research designed to advance the scientific basis of science and innovation policy," which therefore "develops, improves and expands models, analytical tools, data and metrics that can be applied in the science policy decision making process." No single study or single organization will have the broad and lasting impact needed to rationalize policy implementation. But I do believe it is a realistic goal to build a new specialty within the social science community—complete with journals, annual conferences, academic degrees, and chaired professorships—that focuses on the quantitative needs of science policy. This is a good time to encourage such ventures, for at least three reasons.

First, the dramatic influence of information technology on almost every aspect of daily life, from entertainment to global trade, has made

---

[26]Organization for Economic Cooperation and Development, "Science, Technology and Innovation Indicators in a Changing World: Responding to Policy Needs," 2007, www.oecd.org/sti/inno/sciencetechnologyandinnovationindicatorsinachanging worldrespondingtopolicyneeds.htm.

[27]"NSF Workshop on Social Organization of Science and Science Policy," July 13, 2006. For the workshop report, see www.nsf.gov/sbe/scisip/ses_sosp_wksp_rpt.pdf. See also "Science of Science and Innovation Policy: A Prospectus," NSF Social, Behavioral, and Economic Studies Division, September 2006, www.nsf.gov/sbe/scisip/scisip_prospec.pdf.

THE SCIENCE AND ART OF SCIENCE POLICY

it very clear that technical issues will be an important dimension of nearly all future economies. In this context, science and technology policy acquires an unprecedented significance. Post-World War II science policy, at least in the United States, focused on Cold War issues until the late 1980s. The decade of the 1990s was a transition decade. Since the turn of the century all science policy eyes have been on technology-based innovation and how to sustain it. Studies of government science investment strategies have a long history, but the increased demand for economic effectiveness creates a dynamic in which new approaches to science policy studies will flourish.

Second, in the face of rapid global change, old correlations do not have predictive value. The technical workforce today is highly mobile, and information technology has not only dramatically altered the working conditions for technical labor but has also transformed and even eradicated the functions of entire categories of technical personnel. Distributed manufacturing, supply chain management, and outsourcing of ancillary functions have undermined the usefulness of old taxonomies classifying work. The conduct of scientific research itself has been transformed, with extensive laboratory automation, Internet communication and publication, and massive computational and data processing power. The Great Recession of 2008–2009 has, if anything, accelerated the pace of change. We simply must have better tools that do not rely on historical data series. They do not work anymore. Microeconomic reality has inundated macroeconomic tradition with a flood of new behaviors.

Third, the same rapidly advancing technologies that created these new conditions also bring new tools that are particularly empowering for the social sciences. Large databases and complex models are inherent in social science research. The vast articulation of Internet applications makes possible the gathering of socioeconomically relevant data with unprecedented speed and affordability, and access to massive, inexpensive computing power makes it possible to process and visualize data in ways unimagined twenty years ago. New capabilities for direct visualization of large data sets in multiple dimensions may render traditional statistical methods obsolete. A growing community of scientists

228   SCIENCE POLICY UP CLOSE

from many different fields is inventing data mining and data visualization techniques that I believe will transform traditional approaches to analysis and model building. These new tools and opportunities can be invigorating stimuli for all of the social sciences, including the social science of science policy.

If the science of science policy succeeds in establishing itself as a well-defined field and becomes a recognized academic subject, then it is likely to produce three resources that can substantially improve science policy implementation: common, high-quality data resources and interpretive frameworks, a corps of professionals trained in science policy methods and issues, and a network of high-quality communication and discussion that can encompass all science policy stakeholders. Subfields of science policy studies that focus on issues such as energy, climate change, and public health already exist and provide these resources in increasing measure. Climate change controversies notwithstanding, the growth of climate science as an academic field of study, complete with official recognition by international organizations, has manifestly strengthened a consensus for action that did not exist two decades ago. Health policy studies have been around much longer, without which exaggerated or baseless arguments about this or that impact of a given policy would carry much more weight that they do. The existence of a body of peer-reviewed, empirically based analyses makes it possible for news media and Internet-based services to intervene effectively in public policy debate. In the absence of such a resource, the Internet is an advocacy magnifier that adds little substantive value to the public discourse.

Government bears a heavy responsibility to manage the technical resources on which our collective future depends. Some nations, but not ours, provide this management through expert ministries with little public input. Our democracy balances the wasteful entropic forces of public process against the progressive forces of public enlightenment. The inevitable absence of policy discipline in U.S. federal government decision making creates an imperative for some system of public education that fosters rational policy outcomes. The existence of an academic field of science of science policy is a necessary precondi-

THE SCIENCE AND ART OF SCIENCE POLICY   229

tion for such a system. Policies can be formed and carried through rationally only when a sufficient number of men and women follow them in a deep, thoughtful, and open way. Science policy, in its broadest sense, has become so important that it deserves the enduring scrutiny from a profession of its own. This is the promise of the academic discipline of the science of science policy.

Science's Uncertain Authority in Policy

The previous essay dealt with how lack of data and tools threatened the credibility of policy advice; this essay discusses the nature of scientific credibility itself.[28] "My years as White House science advisor made me aware that science has no firm authority in government and public policy," he writes. But this is not a function of this particular administration, for "the authority of science is inferior to statutory authority in a society that operates under the rule of law." Marburger provocatively suggests that scientific authority in government circles falls in the category of what sociologist Max Weber called "charismatic" authority, or based on its perception as "endowed with supernatural, superhuman, or at least specifically exceptional powers or qualities." Scientists find this absurd, Marburger wrote, for "it is precisely because the operation of science does not require charismatic authorities that we should trust it to guide our actions," and it is normal to regard acting against the authority of science as "a mild form of insanity." Still, he continues, unless the authority of science is written into law, the cold truth is that "science is a social phenomenon with no intrinsic authoritative force."

In his unfinished notes for this book, Marburger elaborated on this point, explaining the paradoxical fact that the charismatic authority of science derives from its empirical methods:

Science, scientists, and the science community appear in public discourse as words with definite and stable meanings, but surely this is false. "Scientists," like other people, are multidimensional. Some are smart,

---

[28]John Marburger, "Perspectives: Science's Uncertain Authority in Policy," *Issues in Science and Technology*, Summer 2010, 17–20, issues.org/26-4/p_marburger/.

some are dumb, some wise, and some foolish. They have values, appearances, and beliefs according to accidents of their histories, natured or nurtured. Like other people, they act according to the direction they happen to be looking. They share with all people the desire to do what they want to do and be respected for it. The words themselves, however, signal a power drawn from the effectiveness of empirical methods—methods that require knowledge and skills that are expensive to acquire and therefore not uniformly distributed in any population. As a category, "scientists" inherit social status from owning this label, and therefore a kind of charismatic influence that has encouraged a long history of charlatanism.

Therefore, Marburger concludes, science practitioners and policy makers must continually guard their credibility to be effective, and "science must continually justify itself, explain itself, and proselytize through its charismatic practitioners to gain influence on social events."

—RPC

Scientists view science as the ultimate authority on the laws of the universe, but that authority has no special standing when it comes to the laws of nations. The rigors of the scientific method may be humanity's most reliable approach to attaining rational and objective "truth," but the world's leaders very often follow other routes to policy conclusions. Society's decision makers acknowledge the power of science and invoke its support for their decisions, but they differ greatly from scientists in the way they understand and use science. My years as White House science advisor made me aware that science has no firm authority in government and public policy. Scientists might wish that it were otherwise, but if they want to play an effective role in policy making, they need to understand the political process as it is. A few examples will illustrate my point.

In November 2001, following what were then regarded as incidents of terrorism involving mailed anthrax, Homeland Security advisor Tom Ridge called me seeking urgent advice on what to do with a very large quantity of anthrax-laden U.S. mail. Working with my staff at the White House Office of Science and Technology Policy, we formed an

THE SCIENCE AND ART OF SCIENCE POLICY   231

interagency task group to evaluate and recommend methods to neutralize the spores. The answer we were seeking could not be found in the literature, so we commissioned some research and delivered what was truly "applicable science on demand." We were able to give the U.S. Postal Service precise instructions on how to employ electron-beam irradiation with equipment normally used for food sterilization. Our directions addressed all aspects of the procedure, including the setting for radiation intensity. The Postal Service officials were delighted, and they enthusiastically went to work destroying anthrax—perhaps too enthusiastically. They reported back to us that some of the first batches of mail burst into flame.

We discovered that our guidance, which I would describe as a narrow form of policy advice, was accepted as to method, but not as to degree. Someone surmised that if five on the intensity dial were good, ten would be better. That agent substituted his or her judgment for a well-defined policy recommendation based on careful science and unambiguous data. Much, of course, was at stake. The Postal Service was responsible for delivering mail that would not be lethal. Better to be safe than sorry. When the intensity was throttled back to our recommended level, the treatment worked just fine. You may smile at this minor episode, but it is a relatively benign example of a potentially disastrous behavior. A serious consequence of ignoring expert technical advice occurred in January 1986 when the *Challenger* space shuttle launch rocket failed, killing seven astronauts. The best brief account I know of this tragedy is contained in Edward Tufte's 1997 *Visual Explanations*, which includes a detailed analysis of the manner in which the advice was given. "One day before the flight, the predicted temperature for the launch was 26° to 29° [F]. Concerned that the [O-rings] would not seal at such a cold temperature, the engineers who designed the rocket opposed launching *Challenger* the next day." Their evidence was faxed to the National Aeronautics and Space Administration, where "a high-level NASA official responded that he was 'appalled' by the recommendation not to launch and indicated that the rocket-maker, Morton Thiokol, should reconsider. . . . Other NASA officials pointed out serious weaknesses in the [engineers'] charts. Reassessing the situation after these skeptical responses, the Thiokol managers changed their minds and decided that they now favored launching the next day. They said the evidence presented by the engineers was inconclusive."

Even more was at stake when secret Central Intelligence Agency (CIA) reports to the White House starting in April 2001 advanced the opinion of an analyst—by reasonable standards a well-qualified analyst—that certain aluminum tubes sought by Iraq were likely for use in a nuclear weapons program. That claim was challenged immediately by Department of Energy scientists, probably the world's leading experts in such matters, and later by State Department analysts, who refuted the claim with many facts. The administration nevertheless decided to accept the CIA version in making its case for war. Thanks to a thorough July 2004 report by the Senate Select Committee on Intelligence, the aluminum tubes case is very well documented. This episode is another example of policy actors substituting their subjective judgment in place of a rather clear-cut scientific finding. Did the small group of senior officials who secretly crafted the case for war simply ignore the science? I was not invited to that table, so I cannot speak from direct experience. But I suspect that the process was more complicated than that.

From the evidence that has become available it appears the decision to invade Iraq was based more on a strong feeling among the actors that an invasion was going to be necessary than on a rigorous and systematic investigation that would objectively inform that decision. I will not speculate about the basis for this feeling, but it was very strong. My interest is in how the policy actors in this case regarded science. They were obviously not engaged in a process of scientific discovery. They were attempting to build a case, essentially a legal argument, for an action they believed intuitively to be necessary, and they therefore evaluated the conflicting testimony of credentialed experts from a legal, not a scientific, perspective. The case against the CIA conclusion, although overwhelming from a scientific viewpoint, was nevertheless not absolutely airtight based on material provided to the decision makers. It was reported to the policy-making group by nonscientists who were transmitting summary information in an atmosphere of extreme excite-

THE SCIENCE AND ART OF SCIENCE POLICY   233

ment, stress, and secrecy. I assume that the highly influential CIA briefings on the aluminum tubes did make reference to the Energy Department objections, but this information was transmitted to the decision makers in a way that left a small but real opening for doubt. From a strict legal perspective, seriously limited by the closed and secret nature of the process, that loophole was enough to validate the proposition in their minds as a basis for the desired action.

What is important about these examples is that, as a point of historical fact, the methods of science were weaker than other forces in determining the course of action. The actors had heavy responsibilities, they were working under immense pressure to perform, and the decisions were made within a small circle of people who were not closely familiar with the technical issues. Scientists, and many others, find the disregard of clear technical or scientific advice incomprehensible. Most of us share a belief that the methods of science are the only sure basis for achieving clarity of thought. They are not, unfortunately, the swiftest. The methods of science, as even their articulate champion C. S. Peirce himself observed, do have their disadvantages. Peirce, an eminent logician and the founder of the philosophical school of pragmatism, argued in his famous essays that there are four ways to make our ideas clear and that science is ultimately the only reliable one. However, to quote the *Wikipedia* entry, "Peirce held that, in practical affairs, slow and stumbling ratiocination is often dangerously inferior to instinct, sentiment, and tradition, and that the scientific method is best suited to theoretical research, which in turn should not be bound to the other methods [of settling doubt] and to practical ends." That the physical evidence for Saddam's hypothetical nuclear program was virtually nonexistent, that its significance was appallingly exaggerated in statements by high public officials, and that the consequences of the action it was recruited to justify were cataclysmic, is beside the point. The fact is that although many factors influenced the decision to invade Iraq, science was not one of them, and it is a fair question to ask why not.

To my knowledge, no nation has an official policy that requires its laws or actions to be based on the methods of science. Nor is the aim of science to provide answers to questions of public affairs. That science nevertheless does carry much weight in public affairs must be attributed to something other than the force of law. It is worth asking why advocates of all stripes seek to recruit science to their cause and why we are so offended by actions that "go against science." Studying the source from which science derives its legitimacy may shed some light on conditions under which it is likely to be superseded.

Max Weber, the father of sociology, lists three "pure types of legitimate domination" based on different grounds as follows: (1) "Rational grounds—resting on a belief in the legality of enacted rules and the right of those elevated to authority under such rules to issue commands." This Weber calls legal authority, and he furnishes it with all the bureaucratic trappings of administration and enforcement of what we would call the rule of law. In this case the authorities themselves are rule-bound. (2) "Traditional grounds—resting on an established belief in the sanctity of immemorial traditions and the legitimacy of those exercising authority under them." This is the traditional authority of tribes, patriarchies, and feudal lords. And (3) "Charismatic grounds— resting on devotion to the exceptional sanctity, heroism, or exemplary character of an individual person, and of the normative patterns or order revealed or ordained by him." Weber applies the term *charisma* "to a certain quality of an individual personality by virtue of which he is considered extraordinary and treated as endowed with supernatural, superhuman, or at least specifically exceptional powers or qualities. These are such as are not accessible to the ordinary person."

Weber intended these types to be exhaustive. It is an interesting exercise to attempt to fit the authority of science in society into one or more of these categories. If we admit that science is not sanctioned by law, then of the two remaining choices charismatic authority seems the best match. But to a scientist this is an absurd conclusion. It is precisely because the operation of science does not require charismatic authorities that we should trust it to guide our actions. We tend to accept the authority of science as uniquely representing reality, and to act against it as a mild form of insanity. Experience shows, however, that such insanity is widespread. (Consider only public attitudes toward demonstrably risky behavior such as smoking or texting while driving.) Un-

THE SCIENCE AND ART OF SCIENCE POLICY   235

less it is enforced through legal bureaucratic machinery, the guidance
of science must be accepted voluntarily as a personal policy. Science is
a social phenomenon with no intrinsic authoritative force.

The fact that science has such a good track record, however, endows
its practitioners with a virtue that within the broad social context closely
resembles Weber's "exceptional powers or qualities" that accompany
charismatic authority. And indeed, the public regard for science is
linked in striking ways to its regard for scientists. Contemporary West-
ern culture gives high marks for objectivity, and science, as Peirce com-
pellingly argued, is unique among the ways of making our ideas clear
in arriving at objective, publicly shareable results. In the United States,
at least, there is broad but voluntary public acceptance of science as a
source of authority. Its authority is not mandated, but those who practice it and deliver its results are endowed with charismatic authority.

The National Academies and the National Research Council inherit
this charismatic quality from the status of its members. I was never
more impressed with the power of the Academies and its reports than
in a series of events associated with the development of the proposed
Yucca Mountain nuclear waste repository. The story began with a 1992
law requiring the Environmental Protection Agency (EPA) to base its
safety regulations for the facility on a forthcoming National Research
Council report. When the report appeared in 1995, it implied that science did not preclude drafting radiological safety guidelines extending
over very long times—up to a million years!—related to the half-lives of
certain radioactive components of spent nuclear fuel. Rule making required estimating the impact of potential radiological contamination of
groundwater on populations living in the vicinity of Yucca Mountain
over more than a hundred thousand years. The science of such regulations requires constructing scenarios for both the physical processes of
the storage system and the human population over that time period.
There is no credible and empirically validatable scientific approach for
such long times, and the EPA acknowledged this through a change in
its methodology after 10,000 years. When the regulations were challenged in court, the U.S. Court of Appeals, to my amazement, ruled
that the EPA had not adhered to the letter of the NRC report as required

by law and told EPA to go back to the drawing board. A member of the committee that produced the report, a respected scientist, said that he never expected the report to be used this way. It had become a sacred text. In 2008 both the secretary of energy and the EPA administrator asked my advice on how to proceed, but the issue had passed far beyond the bounds of science. I speculated that in far fewer than a thousand years advances in medical science would have altered completely the significance of hazards such as exposure to low-level ionizing radiation. But such speculations play no role in the formal legal processes of bureaucratic regulation. Yucca Mountain has become a social problem beyond the domain of science.

What emerges from these reflections is that the authority of science is inferior to statutory authority in a society that operates under the rule of law. Its power comes entirely from voluntary acceptance by a large number of individuals, not by any structured consensus that society will be governed by the methods and findings of science. At most, science carries a kind of charismatic power that gives it strength in public affairs but in the final analysis has no force except when embedded in statute. Advocates who view their causes as supported by science work hard to achieve such embedding, and many examples exist of laws and regulations that require consultation with technical expert advisory panels. The Endangered Species Act, for example, "requires the [Fish and Wildlife Service and National Marine Fisheries Service] to make biological decisions based upon the best scientific and commercial data available." Also, "Independent peer review will be solicited . . . to ensure that reviews by recognized experts are incorporated into the review process of rulemakings and recovery plans." The emphasis on "experts" is unavoidable in such regulations, which only sharpens the charismatic aspect of scientific authority. The law typically invokes science through its practitioners, except when adopting specific standards, which are often narrowly prescriptive. Standards, too, however, are established by expert consensus.

At this point the question of the source of scientific authority in public affairs merges with questions about the nature of science itself and its relation to scientists. That society does not automatically accept the authority of science may not come as a surprise. But in my conversations with scientists and science policy makers there is all too often an assumption that somehow science must rule, must trump all other sources of authority. That is a false assumption. Science must continually justify itself, explain itself, and proselytize through its charismatic practitioners to gain influence on social events.

Acknowledgments

This book would not have been written without the assistance and generous support of Carol Marburger, who not only allowed full access to Marburger's notebooks, files, computer, and photographs but also read and commented on the material, making valuable corrections and suggestions. She also gave me permission to quote unpublished material not in the John H. Marburger III Collection in Stony Brook University's Special Collections and University Archives.

Thanks to Jane Yahil, Marburger's literary executor, who diligently tracked down material, read and commented on the text, and provided background context.

Thanks to Kristen Nyitray and Lynn Toscano from the Stony Brook University Special Collections and University Archives for allowing me access to the collection, for help in accessing the collection, and for extending permission to publish material from the John H. Marburger III Collection.

Thanks to historians of science Michael Riordan, Lillian Hoddeson, and Adrienne Kolb, the coauthors of a book in preparation, tentatively titled "Tunnel Visions: The Rise and Fall of the Superconducting Super Collider," who read Chapter 2 and provided helpful comments.

Thanks to Richard Russell, Kathie L. Olsen, and Ann Carlson, with whom Marburger worked at the Office of Science and Technology Policy, for information about his activities there.

Thanks to Patricia Flood in the Information Technology Division at Brookhaven National Laboratory for help retrieving some of Marburger's materials.

Thanks to Michael G. Fisher, the Executive Editor for Science and Medicine at Harvard University Press, for recognizing the value of this book; to Lauren

Esdaile of Harvard University Press for helping to shepherd it through; and to copyeditor Trish Watson and production editor Deborah Grahame-Smith. Jyotsna Jeyapaul and Travis Holloway helped with the index; Holloway also helped with the editing.

"Man in the News; A Mediator for Shoreham Panel: John Harmen Marburger III" in Chapter 1 is from the *New York Times*, December 15, 1983, and is used with permission.

An excerpt from Chapter 2, "The Superconducting Super Collider and the Collider Decade," appeared in *Physics in Perspective*, 16:2, and is used with permission.

The address on the future of the Brookhaven National Laboratory in Chapter 3 and the AAAS addresses in Chapters 4 and 5 are from the John H. Marburger III Collection, Special Collections and University Archives, Stony Brook University Libraries, and are used with permission.

The *Newsday* interview with John Marburger, "It Takes More than Science to Do Science," in Chapter 3 appeared in *Newsday*, December 30, 1997, and is used with permission.

"Why Policy Implementation Needs a Science of Science Policy" in Chapter 6 is from *The Science of Science Policy: A Handbook*, edited by Kaye Husbands Fealing, Julia I. Lane, John H. Marburger III, and Stephanie S. Shipp, copyright © 2011 by the Board of Trustees of the Leland Stanford Jr. University. All rights reserved. Used with the permission of Stanford University Press, www .sup.org.

"Science's Uncertain Authority in Policy" in Chapter 6 appeared in *Issues in Science and Technology*, Summer 2010, 17–20, and is reprinted with permission of the University of Texas at Dallas, Richardson, Texas.

Index

Accelerators, 30–72, 79, 88, 93, 107–113; Alternating Gradient Synchrotron (AGS), 39, 76, 112; Deutsches Elektronen Synchrotron (DESY), 37; Isabelle/CBA, 33–39, 43, 50, 75; Large Hadron Collider (LHC), 76, 116; Main Ring (Fermilab), 31, 33–34, 37, 39, 43–44, 48, 50, 54, 59, 60–63, 72, 79, 144; Proton Synchrotron (PS), 33, 37, 39; Relativistic Heavy Ion Collider (RHIC), 6, 72, 74, 75, 76, 87, 88, 107–113, 116; Stanford Linear Accelerator Center (SLAC), 37, 52, 79, 191; Superconducting Super Collider (SSC), 5, 30, 31, 32, 38–72, 75, 76, 79, 173, 174, 239; Super Proton Synchrotron (SPS), 33, 37; Tevatron, 34, 43
Advanced Energy Initiative, 170
Advice, 3, 9–10, 15, 23, 39, 63, 119–128, 133–134, 145, 151, 204, 206–208, 221, 224, 229, 230–236
Advocacy, 3, 35, 62, 121–125, 158–165, 168, 181, 207–221, 228
"Allocating Federal Funds for Science and Technology" (report), 161, 175
Alternating Gradient Synchrotron (AGS), 39, 76, 112
Aluminum tubes, as justification for Iraq invasion, 232–233
"America Creating Opportunities to Meaningfully Promote Excellence in Technology, Education, and Science Act of 2007" (America COMPETES Act), 170, 193, 196, 200, 218
American Association for the Advancement of Science (AAAS), 125–126, 135, 144,

146, 156–159, 168–170, 179–202, 208–210, 215, 224–225
American Competitiveness Initiative (ACI), 8, 123, 169–188, 193, 196, 200–201, 218–220
American Recovery and Reinvestment Act (ARRA, 2009), 218
Anthrax, 149, 230–231
Apollo program, 147, 176
Aronson, Samuel, 116
Associated Universities, Inc. (AUI), 33, 35, 40–41, 73, 76, 78, 82, 91
Atlantic, The, 36
Atomic Energy Commission, U.S. (AEC), 33, 39–42, 48–49, 53
Augustine, Norman, 178
Augustine Committee, 216
Authority, 9, 11, 24, 50, 67, 172, 205, 211, 215, 217, 229–237
Axelrod, David, 29

Balance, 3, 12, 35, 62, 76–79, 126, 129–132, 200
Baltimore, David, 144
Basic research, 35, 73, 79, 129, 148, 158, 163, 165, 167, 175, 183–184, 187–188, 191, 201, 209, 214, 219
Battelle Memorial Institute, 73, 77, 90
Bayh-Dole Act, 191
Benchmarks, 158, 164–165, 179, 189, 220
Bernanke, Ben, 208–209
Berners-Lee, Timothy, 174
Bioterrorism, 127, 144, 149
Black holes, 72, 107, 109
Blue Sky II (conference), 225–226

Blume, Martin, 52
Branscomb, Lewis, 128
Bromley, Allan, 170, 172
Brookhaven Graphite Research Reactor
 (BGRR), 80, 82
Brookhaven National Laboratory (BNL), 1, 2,
 6, 9, 16, 23, 27, 29, 33–41, 44, 48, 50, 52,
 73–118, 146, 239
Brookhaven Science Associates (BSA), 73,
 77, 83, 99
Brown, George, 174
Burden of representation, 7, 85, 202
Bureau of Budget (BOB), 39, 41
Burstein, Karen S., 29
Bush, George H. W., 52, 68
Bush, George W., 7, 115, 119, 121–122, 144,
 146, 156, 169, 180, 200–204, 212
Bush, Vannevar, 214
Busza, Wit, 108

California Institute of Technology, 144
Campo, Leon J., 27, 29
Carey, Hugh, 15
Carlson, Ann, 208–209, 239
Carney, William, 35
Carson, Rachel, 69
Carter, Jimmy, 31, 49–50, 69
Catacosinos, William, 24
Center for Risk Science and Communi-
 cation, 116
Center for Science and Technology Policy
 Research, 156
Central Design Group (CDG), 46, 58
Central Intelligence Agency (CIA),
 232–233
CERN, 33–34, 37, 39, 116, 174
Challenger space shuttle, 231
Charisma, 11, 229–237
Chernobyl accident, 26, 67–68, 100, 102
Chronicle of Higher Education, 56–57,
 136
Cipriano, Joseph, 67
Clean Air Act, 69
Clean Water Act, 69–70
Climate change, 120, 125, 144, 160, 180–181,
 188, 197, 218, 228
Clinton, William J., 31, 176
Clough, Wayne, 176
Cohalan, Peter, 15, 18, 24
Cold War, 31, 68, 75, 79–80, 155, 171, 174,
 201, 215, 219, 227
Colliding Beam Accelerator (CBA), 38, 75

Combating Terrorism Technical Support
 Working Group, 128
Commons, 192–193, 197–200; "Tragedy of
 the Commons" (article), 193, 197, 199
Community Advisory Council, 115, 118
CONDOR, 141–143
Congress, U.S., 5, 30–53 *passim*, 80, 120,
 124, 136, 139, 147–150, 159, 165–166,
 169–202 *passim*, 213, 216–221, 226
*Constructing Reality: Quantum Theory and
 Particle Physics* (book), 9, 207
Contact (movie), 86
Cozzens, Susan, 210
Crease, Robert P., 1, 33–34, 38–39, 48, 72, 77
Credibility, 7, 11, 51, 82, 202, 207, 224,
 229–230
Cuomo, Mario P., 4, 5, 13–15, 19, 25–28,
 77, 122

Dale, Shana, 142
Daniels, Mitch, 124, 132, 210
Dar, Arnon, 112
D'Ascoli, Jeanne, 116
Decker, James, 64, 67
Deming, W. E., Plan–Do–Check–Act
 cycle, 54
Department of Defense (DOD), 128, 170,
 174, 177, 184–187, 201, 215, 217
Department of Energy (DOE), 6, 30–41, 45,
 67, 73, 81– 83, 99, 102, 106, 153, 174,
 176–177, 183, 199, 215, 217, 232; High
 Energy Physics Advisory Panel
 (HEPAP), 38–39, 45; Office of
 Environment, Safety, and Health,
 82–83; Office of Science, 61, 153,
 176–177, 183
De Rujula, Alvaro, 112
Deutch, John, 51, 61
Deutsches Elektronen Synchrotron
 (DESY), 37
"Dilemma of the Golden Age" (speech), 172
Dircks, William J., 29
Dirksen, Everett, 41
"Discovery-oriented" versus "issue-
 oriented" science, 126, 131–132
"Does Science Policy Exist? And If So–Does
 It Matter?" (article), 156, 195, 208
Douglas, Mary, 20
Drama of the Commons (survey), 197
Drell, Sidney, 144
Dullea, Hank, 14–15
Duncan, Charles, 49

INDEX

243

Earmarks, 161–163, 178–179, 184–187, 198, 217
Ehlers, Vernon, 80, 83, 175, 199
Eisenhower, Dwight, 119, 213
End of Science, The (book), 89
Endangered Species Act, 69, 236
Energy, 3–4, 7, 15–16, 27, 29, 33–38, 39, 41, 46, 49–50, 56, 61–112 passim, 127, 153, 160, 173, 178, 180–181, 185–186, 201, 218, 226, 228, 236
Energy Research and Development Agency (ERDA), 49
Enhanced Border Security and Visa Entry Reform Act of 2002, 137
Environmental Impact Statement (EIS), 91
Environmental Protection Agency (EPA), 69–70, 99, 185–186, 235–236
Environment Health & Safety (EH&S), 70–72, 83–84, 118
Executive time, 3–4, 6, 73, 85, 122
Exogenous factors, 48

Federal Acquisition Regulations (FAR), 58
Federal Bureau of Investigation (FBI), 70, 140–141
Federal science and technology (FS&T; budget category), 161–162
Federal science budget, 157, 189, 203, 208
Fermilab (FNAL–Fermi National Accelerator Laboratory), 31–63 passim, 72, 79, 144; Main Injector, 62; Main Ring, 39, 43, 50, 54; Tevatron, 34, 43
Fink Committee, 149
Fitch, Val, 35
Forbes, Michael, 88
Forsythe, Dall, 50, 53, 72
Forum on Science and Technology Policy, AAAS, 144, 146, 157, 169, 179, 192
Fox, Marye Anne, 183
Frontiers of Electronics (TV show), 2

Gardner, Howard, 8, 145
Gates, Robert, 201
General Accountability Office (GAO), 55–57, 68, 217
Geological Survey, U.S. (USGS), 185–186, 200
George C. Marshall Institute, 203
Gingrich, Newt, 80, 175
Global Change Research Program (GCRP), 214

Gore, Al, 180–181
Government Performance and Results Act (GPRA), 50, 53, 72
Greider, William, 36
Gross Domestic Product (GDP), 147, 163–164, 188–189, 207
Gross National Product (GNP), 7, 172
Groves, Leslie R., 48

Hardin, Garrett, 193, 197, 199
Harrison, Marge, 29
Harvard International Review, 197
Heinz, Ulrich, 112
Herrington, John, 67
High Energy Physics Advisory Panel (HEPAP), 38–39, 45
High Flux Beam Reactor (HFBR), 74–107 passim
Hoddeson, Lillian, 30, 39, 42, 44, 47, 58, 59, 60, 72, 239
Homeland Security, 127–149 passim, 182, 184, 230; Act, 137
Horgan, John, 89
Hornig, Donald, 40
Hosmer, Craig, 41
House Science Committee, 65, 137–138, 174, 194
Hunter, Robert O., 64, 67
Hussein, Saddam, 233
Hut, Piet, 110
Hydrogen Fuel Initiative, 153

Illegal Immigration Reform and Immigrant Responsibility Act, 1996, 139
Immigration and Nationality Act (INA), 137, 139–140
Inconvenient Truth, An (movie), 180
Information technology, 131, 148, 154, 160, 163, 174, 211, 226–227
Institut Laue-Langevin (ILL), 96
Interagency Panel on Advanced Science and Security (IPASS), 138, 143
Intergovernmental Panel on Climate Change (IPCC), 181
International Space Station (ISS), 174
International Thermonuclear Experimental Reactor (ITER), 201
Iraq, U.S. invasion of, 7, 232, 233
Isabelle/Colliding Beam Accelerator (CBA), 33–39, 43, 50
Issues in Science and Technology, 156, 195, 208, 229, 240

Jack Daniels memo, 124, 210
Jackson, David, 47
Jacobs, Janice, 137, 141
Jaffe, Robert, 108
Jeffrey, William, 142
Johnson, Lyndon, 31, 41, 50
Joint Committee on Atomic Energy
    (JCAE), 40
Joint Telecommunications Resources
    Board, 121

Kahn, Alfred E., 28–29
Kass, Leon R., 121
Kavli Foundation, 191
Keyworth, George (Jay), 52
Klausner, Richard, 128
Klein, John V. N., 24
Knapp, Ed, 51, 52, 60, 62–63, 65
Koizumi, Kei, 198
Kolb, Adrienne, 30, 44, 58–59, 72, 239
Koppelman, Lee R., 15
Koshland Museum of Science, 180
Kouts, Herbert J., 27, 29

Laboratories: Brookhaven National
    Laboratory (BNL), 1, 2, 6, 9, 16, 23, 27,
    29, 33–41, 44, 48, 50, 52, 73–118, 146,
    239; CERN, 33–34, 37–39, 116, 174;
    Deutsches Elektronen Synchrotron
    (DESY), 37; Fermilab (FNAL–Fermi
    National Accelerator Laboratory), 31–63
    passim, 72, 79, 144; Lawrence Berkeley
    Laboratory (LBL), 31, 39; Stanford Linear
    Accelerator Center (SLAC), 37, 52, 79, 191
Large Hadron Collider (LHC), 76, 116
LaRocca, James, 16
Lawrence, Ernest, 42, 48
Lawrence Berkeley Laboratory (LBL), 31, 39
Lederman, Leon, 37–38, 44, 50, 60, 144
"Legalistic mechanisms of accountability,"
    78–79
Leno, Jay, 122
Long Island Lighting Company (LILCO):
    13–25
Long Island Power Authority (LIPA), 24–25
Lynch, Margaret, 116

Main Injector (Fermilab), 62
Main Ring (Fermilab), 39, 43, 50, 54
Making Physics: A Biography of Brookhaven
    National Laboratory, 1946–1972 (book),
    33, 72

"Making the Nation Safer" (report), 147
Management and Operating (M&O)
    contracts, 58, 61, 70
Manhattan Project, 30, 33, 47, 49
Manhattan Projectitis, 31
Mann, Charles C., 77
MANTIS, 141–143
Marburger, Alexander (son), 29
Marburger, Carol, 239
Marburger, John (son), 29
Marks, Paul Allen, 29
McDaniel, Boyce, 59–60, 65
McDonald, Kim, 56–57
McLoughlin, David, 29
McMillan, Edwin, 42
"Measuring Research and Development
    Expenditures in the U.S. Economy"
    (report), 167
Moore's law, 165
Morton Thiokol, 231

Nanotechnology, 131, 144, 150–151, 154, 160,
    165–166
Nanotechnology Research and Devel-
    opment Act, 150
Narrative, 5–6, 10, 166, 203, 207
National Academy of Sciences (NAS), 40,
    67–68, 72, 128, 151, 156, 169, 172, 180,
    218, 221
National Aeronautics and Space Admin-
    istration (NASA), 147, 153, 176, 185–186,
    214–219, 231
National Environmental Policy Act (NEPA),
    69
National Institute for Standards and
    Technology (NIST), 177–178, 183,
    185–186, 200–201
National Institutes of Health (NIH), 10,
    130–131, 152, 169, 176, 185–186, 189–190,
    199, 201, 215, 217, 219
National Nanotechnology Initiative (NNI),
    165, 176, 214, 216
National Oceanic and Atmosphere
    Administration (NOAA), 185–186, 200
National Public Radio (NPR), 145
National Research Council (NRC), 133, 151,
    161, 167, 197, 221, 235
National Science Advisory Board for
    Biosecurity (NSABB), 149, 183
National Science and Technology Council
    (NSTC), 127, 208–209, 214, 216
National Science Board (NSB), 222–224

INDEX   245

National Science Foundation (NSF), 9–10,
    36, 130, 153, 161, 164, 166–167, 170, 176,
    183, 185–186, 189, 199, 210, 214–215, 217,
    223, 226
National Security Council (NSC), 123
Networking and Information Technology
    Research and Development (NITRD)
    program, 176, 214
Newsday, 12, 25, 85–86, 240
New York State, 4, 13–14, 16, 26, 33, 35, 77;
    Board of Regents, 33; 1st congressional
    district, 35
New York Times Magazine, 205
Nixon, Richard, 31, 49–50
Nuclear Fear (book), 102
Nuclear Regulatory Commission (NRC), 19,
    24–25, 29, 49, 69, 167, 175, 221, 235

Office of Energy Research (OER), 52–53,
    64, 67
Office of Homeland Security, 127, 142
Office of Management and Budget (OMB),
    36, 39, 121, 123–124, 127, 133, 163,
    178–179, 185–187, 198, 210, 215, 217
Office of Science, Department of Energy,
    61, 153, 176–177, 183
Office of Science and Technology Policy
    (OSTP), 7, 52, 119–129, 136, 142, 145,
    150–151, 156, 166, 179–182, 186, 194,
    203–204, 208, 212–217, 220, 230, 239
Office of the Superconducting Super
    Collider (OSSC), 61–66
Olsen, Kathie, 208–209, 211, 239
"On Management" (JHM article), 3–4
Organization for Economic Cooperation
    and Development (OECD), 164, 189,
    209, 225; Global Science Forum, 209

Panofsky, Wolfgang, 52–53, 60, 65
Patriot Act, 2003, 139
Peirce, Charles S., 233, 235
Peña, Federico, 73, 75–76, 82–83,
    87–88, 91
Peoples, John, 44
Pewitt, N. Douglas, 52–53
Physics Today, 52–53
Physics World, 192, 203
Pielke, Roger, Jr., 156, 158, 163
Postal Service, U.S., 231
Presidential science advisor, 1, 8, 10, 40, 52,
    119–122, 125, 144, 146, 156–157, 170, 192,
    203, 207–208

President's Council of Advisors on Science
    and Technology (PCAST), 127, 136, 151,
    155, 166, 176–177, 190
President's Council on Bioethics, 121
Press, Frank, 156, 161, 164, 169, 172–177, 221
Principle of Assurance, 6, 10, 32, 49, 53–54,
    70, 207
Prioritization, 3, 124, 126, 162, 175, 179
"Proposed Strategies for Minimizing the
    Potential Misuse of Life Sciences
    Research" (draft report), 183
Proton Synchrotron (PS), 33, 37, 39
Public Health Security and Bioterrorism
    Preparedness and Response Act,
    2002, 149

Quigg, Chris, 47

Ramsey, Norman, 39, 42, 60
Reactors, nuclear, 20, 67, 91–107, 132
Reagan, Ronald, 35–36, 51–52, 64, 69, 75,
    172, 219
Rees, Martin J., 110
Reference Design Study (RDS), 45–46
Relativistic Heavy Ion Collider (RHIC), 6,
    72, 74, 75, 76, 87, 88, 107–113, 116
Research America, 219
Research and development (R&D), 39, 127,
    135, 147–148, 154–163, 167, 176, 189, 209,
    219
Resource Conservation and Recovery Act,
    69–70
"Restoring Scientific Integrity in Policy
    Making" (document), 8, 144
Rezendes, Victor, 55–56
Richardson, William, 91, 105–107, 240
Ridge, Thomas, 230
Riordan, Michael, 30, 47, 60, 239
"Rising above the Gathering Storm:
    Energizing and Employing America for
    a Brighter Economic Future" (report),
    177, 218
Risk, 6, 20–22, 32, 36, 54, 55, 57, 58, 79,
    91–118, 129, 149, 169, 175, 177
Risk and Culture (book), 20
Rockefeller, Nelson, 63
Rocky Flats (weapons facility), 70
Ronan, William J., 29
Roosevelt Room (White House), 123
Roth, Patricia, 18
Rubbia, Carlo, 37
Russell, Richard, 204, 239

Safe Drinking Water Act, 69
Salgado, Joseph, 61
Samios, Nicholas, 35, 38, 73, 76
Sandweiss, Jack, 108
Sarewitz, Daniel, 7, 156
Schlesinger, James, 49, 51
Schwitters, Roy, 60–61, 65
Science (magazine), 80, 158, 197, 208
Science and Engineering Indicators, 161,
    164, 166, 223
Science and Industry (book), 1
"Science-based" and "issues-based" policy,
    126
Science of Science and Innovation Policy
    (SciSIP), 226
Science of Science Policy, A Handbook, The,
    11, 207, 212, 240
Scientific American (magazine), 107
Seaborg, Glenn, 39–42, 49–50
*Second Creation: Makers of the Revolution in
    Twentieth-Century Physics, The* (book),
    77
Security, 8, 79–80, 121, 134–150, 157, 159, 171,
    176, 182, 199, 222
Seitz, Frederick, 40
Senate Select Committee on Intelligence,
    232
Senior staff meeting, 123
September 11, 2001 terrorist attacks, 146,
    182, 221
Sheridan, Tom, 106–107
Shoreham Nuclear Power Plant, 4, 13, 35;
    Shoreham Nuclear Power Plant
    Commission, 4
Silent Spring (book), 69
Simon, Robert M., 56, 65
Siskin, Ed, 56–57
Small Business Innovation Research
    (SBIR), 199
Snowmass, Colorado, 36, 38, 45
"Social Dimensions of Science" (course), 9
Social sciences, 126, 133–134, 153, 208,
    227–228
Soviet Union, 31, 48, 80, 100
Spinoza, Baruch, 10
Sputnik, 159, 174, 219
Standard Model, 33, 37, 48, 77, 111
Stanford Linear Accelerator Center (SLAC),
    37, 52, 79, 191
State of the Union address, 8, 169
Stem cell research, 7, 121, 202, 205
Stevenson-Wydler Act, 191

Stockman, David, 35–36
Stony Brook University (SBU), 1, 3, 9, 12–15,
    24, 33, 62, 73, 119, 123–125, 145, 239
Strangelets, 107, 110–113
Strange matter, 110–113
Strategic Defense Initiative (SDI), 219
Student and Exchange Visitor Information
    System (SEVIS), 137, 139, 149
Suffolk County, 14–15, 19, 21–22, 75, 77, 81
Suffolk County Water Authority
    (SCWA), 94
Sunday Times (London), 107
Superconducting Super Collider (SSC), 5,
    30, 31, 32, 38–72, 75, 76, 79, 173, 174,
    239; Laboratory (SSCL), 44, 51, 54–67;
    Office of (OSSC), 61–66
Superfund Act, 69, 99
Sverdrup Corporation, 64

Task Force on the Future of American
    Innovation, 219
Technology Alert List (TAL), 141, 143
Temple, Edward, 65
Tevatron (accelerator), 34, 43
Three Mile Island nuclear facility, 17
Tiger Teams, 70
Tigner, Maury, 45–47, 59– 61
Ting, Samuel, 35
Toll, John S., 62–63
Tritium, 74, 81–82, 87, 90, 97–100, 116
Trivelpiece, Alvin, 64, 172
Trust, 10–11, 20–21, 85, 102, 121, 123, 229, 234
Tufte, Edward, 231
Tuskegee University, 170

Union of Concerned Scientists, 8, 144, 205
Universities Research Association (URA), 1,
    5, 30–32, 40, 51, 169, 173; Board of
    Overseers (BOO), 58–65; Board of
    Trustees (BOT), 5, 30, 40–41, 44, 59,
    62, 65, 240; Council of Presidents
    (COP), 5, 30, 41–44, 51, 59, 61
University of Southern California (USC), 2,
    12, 23, 63
"Unlocking Our Future: Toward a New
    National Science Policy" (report), 175

Vacuum instability, 110
Van der Meer, Simon, 37
Varmus, Harold, 175, 219
Visas, 138–146 *passim*, 150, 222–223
Visual Explanations (book), 231

War against terrorism, 126–129, 160
Water Pollution Control Act, 69
Watkins, James D., 65, 67–72
Weart, Spencer, 102
Weber, Max, 11, 229, 234–235
White House, 1, 7, 9, 36, 39, 41, 119–123, 145, 157, 178, 181, 185, 193–194, 214–216, 229–232
White House Council on Environmental Quality, 69
Wilczek, Frank, 107–108
Wildavsky, Aaron, 20
Wilson, Hugh, 29
Wilson, Robert R., 42–43, 45, 50, 54, 69
Winerip, Michael, 26

Wojcicki, Stanley, 38, 44, 47, 60, 64–65, 67, 72
Women in Science and Engineering (WISE), 9
Workforce, 8, 126, 134, 140, 157, 163, 168, 171, 200, 220–227
World Summit on the Information Society (WSIS), 121
World War II, 30, 34, 80, 175, 219, 227

Yale University, 33, 108, 172
Yang, C. N., 34–35
Yucca Mountain nuclear waste repository, 235

Zero-based budgeting, 49–50